Student Edition

Eureka Math
Geometry
Modules 1 & 2

Special thanks go to the Gordan A. Cain Center and to the Department of Mathematics at Louisiana State University for their support in the development of *Eureka Math*.

Published by Great Minds

Copyright © 2015 Great Minds. All rights reserved. No part of this work may be reproduced or used in any form or by any means — graphic, electronic, or mechanical, including photocopying or information storage and retrieval systems — without written permission from the copyright holder. "Great Minds" and "Eureka Math" are registered trademarks of Great Minds.

Printed in the U.S.A.
This book may be purchased from the publisher at eureka-math.org
10 9 8 7 6 5 4 3
ISBN 978-1-63255-327-0

Lesson 1: Construct an Equilateral Triangle

Classwork

Opening Exercise

Joe and Marty are in the park playing catch. Tony joins them, and the boys want to stand so that the distance between any two of them is the same. Where do they stand?

How do they figure this out precisely? What tool or tools could they use?

Fill in the blanks below as each term is discussed:

a. _____ The _____ between points A and B is the set consisting of A, B, and all points on the line AB between A and B.

b. _____ A segment from the center of a circle to a point on the circle

c. _____ Given a point C in the plane and a number $r > 0$, the _____ with center C and radius r is the set of all points in the plane that are distance r from point C.

Note that because a circle is defined in terms of a distance, r, we often use a distance when naming the radius (e.g., "radius AB"). However, we may also refer to the specific segment, as in "radius $\overline{AB}$."

Example 1: Sitting Cats

You need a compass and a straightedge.

Margie has three cats. She has heard that cats in a room position themselves at equal distances from one another and wants to test that theory. Margie notices that Simon, her tabby cat, is in the center of her bed (at S), while JoJo, her Siamese, is lying on her desk chair (at J). If the theory is true, where will she find Mack, her calico cat? Use the scale drawing of Margie's room shown below, together with (only) a compass and straightedge. Place an M where Mack will be if the theory is true.

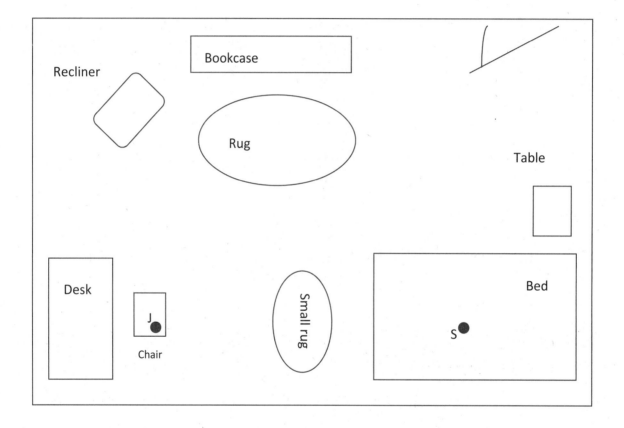

Lesson 1: Construct an Equilateral Triangle

Mathematical Modeling Exercise: Euclid, Proposition 1

Let's see how Euclid approached this problem. Look at his first proposition, and compare his steps with yours.

Proposition 1

To construct an equilateral triangle on a given finite straight-line.

In this margin, compare your steps with Euclid's.

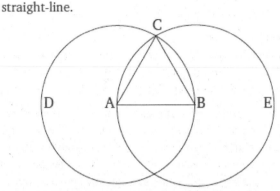

Let AB be the given finite straight-line.
So it is required to construct an equilateral triangle on the straight-line AB.
Let the circle BCD with center A and radius AB have been drawn [Post. 3], and again let the circle ACE with center B and radius BA have been drawn [Post. 3]. And let the straight-lines CA and CB have been joined from the point C, where the circles cut one another,† to the points A and B (respectively) [Post. 1].
And since the point A is the center of the circle CDB, AC is equal to AB [Def. 1.15]. Again, since the point B is the center of the circle CAE, BC is equal to BA [Def. 1.15]. But CA was also shown (to be) equal to AB. Thus, CA and CB are each equal to AB. But things equal to the same thing are also equal to one another [C.N. 1]. Thus, CA is also equal to CB. Thus, the three (straight-lines) CA, AB, and BC are equal to one another.
Thus, the triangle ABC is equilateral, and has been constructed on the given finite straight-line AB. (Which is) the very thing it was required to do.

Lesson 1: Construct an Equilateral Triangle

Geometry Assumptions

In geometry, as in most fields, there are specific facts and definitions that we assume to be true. In any logical system, it helps to identify these assumptions as early as possible since the correctness of any proof hinges upon the truth of our assumptions. For example, in Proposition 1, when Euclid says, "Let AB be the given finite straight line," he assumed that, *given any two distinct points, there is exactly one line that contains them*. Of course, that assumes we have two points! It is best if we assume there are points in the plane as well: *Every plane contains at least three noncollinear points*.

Euclid continued on to show that the measures of each of the three sides of his triangle are equal. It makes sense to discuss the measure of a segment in terms of distance. *To every pair of points A and B, there corresponds a real number $\text{dist}(A,B) \geq 0$, called the distance from A to B.* Since the distance from A to B is equal to the distance from B to A, we can interchange A and B: $\text{dist}(A,B) = \text{dist}(B,A)$. Also, A *and* B *coincide if and only if* $\text{dist}(A,B) = 0$.

Using distance, we can also assume that *every line has a coordinate system*, which just means that we can think of any line in the plane as a number line. Here's how: Given a line, l, pick a point A on l to be "0," and find the two points B and C such that $\text{dist}(A,B) = \text{dist}(A,C) = 1$. Label one of these points to be 1 (say point B), which means the other point C corresponds to -1. Every other point on the line then corresponds to a real number determined by the (positive or negative) distance between 0 and the point. In particular, if after placing a coordinate system on a line, if a point R corresponds to the number r, and a point S corresponds to the number s, then the distance from R to S is $\text{dist}(R,S) = |r - s|$.

History of Geometry: Examine the site http://geomhistory.com/home.html to see how geometry developed over time.

Relevant Vocabulary

GEOMETRIC CONSTRUCTION: A *geometric construction* is a set of instructions for drawing points, lines, circles, and figures in the plane.

The two most basic types of instructions are the following:

1. Given any two points A and B, a straightedge can be used to draw the line AB or segment AB.
2. Given any two points C and B, use a compass to draw the circle that has its center at C that passes through B. (Abbreviation: Draw circle C: center C, radius CB.)

Constructions also include steps in which the points where lines or circles intersect are selected and labeled. (Abbreviation: Mark the point of intersection of the line AB and line PQ by X, etc.)

FIGURE: A (two-dimensional) *figure* is a set of points in a plane.

Usually the term figure refers to certain common shapes such as triangle, square, rectangle, etc. However, the definition is broad enough to include any set of points, so a triangle with a line segment sticking out of it is also a figure.

EQUILATERAL TRIANGLE: An *equilateral triangle* is a triangle with all sides of equal length.

COLLINEAR: Three or more points are *collinear* if there is a line containing all of the points; otherwise, the points are noncollinear.

Lesson 1: Construct an Equilateral Triangle

LENGTH OF A SEGMENT: The *length of* $\overline{AB}$ is the distance from A to B and is denoted AB. Thus, $AB = \text{dist}(A, B)$.

In this course, you have to write about distances between points and lengths of segments in many, if not most, Problem Sets. Instead of writing $\text{dist}(A, B)$ all of the time, which is a rather long and awkward notation, we instead use the much simpler notation AB for both distance and length of segments. Even though the notation always makes the meaning of each statement clear, it is worthwhile to consider the context of the statement to ensure correct usage.

Here are some examples:

- $\overleftrightarrow{AB}$ intersects... $\overleftrightarrow{AB}$ refers to a line.
- $AB + BC = AC$ Only numbers can be added, and AB is a length or distance.
- Find $\overline{AB}$ so that $\overline{AB} \parallel \overline{CD}$.   Only figures can be parallel, and $\overline{AB}$ is a segment.
- $AB = 6$ AB refers to the length of $\overline{AB}$ or the distance from A to B.

Here are the standard notations for segments, lines, rays, distances, and lengths:

- A ray with vertex A that contains the point B: $\overrightarrow{AB}$ or "ray AB"
- A line that contains points A and B: $\overleftrightarrow{AB}$ or "line AB"
- A segment with endpoints A and B: $\overline{AB}$ or "segment AB"
- The length of $\overline{AB}$: AB
- The distance from A to B: $\text{dist}(A, B)$ or AB

COORDINATE SYSTEM ON A LINE: Given a line l, a *coordinate system on* l is a correspondence between the points on the line and the real numbers such that: (i) to every point on l, there corresponds exactly one real number; (ii) to every real number, there corresponds exactly one point of l; (iii) the distance between two distinct points on l is equal to the absolute value of the difference of the corresponding numbers.

Lesson 1: Construct an Equilateral Triangle

Problem Set

1. Write a clear set of steps for the construction of an equilateral triangle. Use Euclid's Proposition 1 as a guide.

2. Suppose two circles are constructed using the following instructions:

   Draw circle: center A, radius $\overline{AB}$.

   Draw circle: center C, radius $\overline{CD}$.

   Under what conditions (in terms of distances AB, CD, AC) do the circles have

   a. One point in common?
   b. No points in common?
   c. Two points in common?
   d. More than two points in common? Why?

3. You need a compass and straightedge.

   Cedar City boasts two city parks and is in the process of designing a third. The planning committee would like all three parks to be equidistant from one another to better serve the community. A sketch of the city appears below, with the centers of the existing parks labeled as P_1 and P_2. Identify two possible locations for the third park, and label them as P_{3a} and P_{3b} on the map. Clearly and precisely list the mathematical steps used to determine each of the two potential locations.

Lesson 2: Construct an Equilateral Triangle

Classwork

Opening Exercise

You need a compass, a straightedge, and another student's Problem Set.

Directions:

Follow the directions of another student's Problem Set write-up to construct an equilateral triangle.

- What kinds of problems did you have as you followed your classmate's directions?
- Think about ways to avoid these problems. What criteria or expectations for writing steps in constructions should be included in a rubric for evaluating your writing? List at least three criteria.

_____

_____

_____

_____

_____

Exploratory Challenge 1

You need a compass and a straightedge.

Using the skills you have practiced, construct **three** equilateral triangles, where the first and second triangles share a common side and the second and third triangles share a common side. Clearly and precisely list the steps needed to accomplish this construction.

Switch your list of steps with a partner, and complete the construction according to your partner's steps. Revise your drawing and list of steps as needed.

Construct three equilateral triangles here:

Lesson 2: Construct an Equilateral Triangle

Lesson 2

Exploratory Challenge 2

On a separate piece of paper, use the skills you have developed in this lesson construct a **regular hexagon**. Clearly and precisely list the steps needed to accomplish this construction. Compare your results with a partner, and revise your drawing and list of steps as needed.

Can you repeat the construction of a hexagon until the entire sheet is covered in hexagons (except the edges are partial hexagons)?

_____

_____

_____

_____

_____

Lesson 2: Construct an Equilateral Triangle

Problem Set

Why are *circles* so important to these constructions? Write out a concise explanation of the importance of circles in creating equilateral triangles. Why did Euclid use *circles* to create his equilateral triangles in Proposition 1? How does construction of a circle ensure that all relevant segments are of equal length?

_____

_____

_____

_____

_____

Lesson 3: Copy and Bisect an Angle

Classwork

Opening Exercise

In the following figure, circles have been constructed so that the endpoints of the diameter of each circle coincide with the endpoints of each segment of the equilateral triangle.

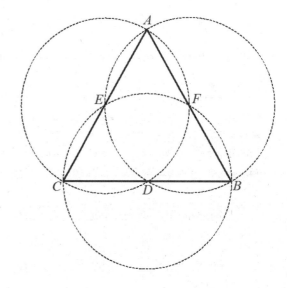

a. What is special about points D, E, and F? Explain how this can be confirmed with the use of a compass.

b. Draw $\overline{DE}$, $\overline{EF}$, and $\overline{FD}$. What kind of triangle must $\triangle DEF$ be?

c. What is special about the four triangles within $\triangle ABC$?

d. How many times greater is the area of $\triangle ABC$ than the area of $\triangle CDE$?

A STORY OF FUNCTIONS

Lesson 3 M1

GEOMETRY

Discussion

Define the terms **angle**, **interior of an angle**, and **angle bisector**.

ANGLE: An *angle* is _____

_____

_____

_____

INTERIOR: The *interior of* $\angle BAC$ is the set of points in the intersection of the half-plane of $\overleftrightarrow{AC}$ that contains B and the half-plane of $\overleftrightarrow{AB}$ that contains C. The interior is easy to identify because it is always the "smaller" region of the two regions defined by the angle (the region that is convex). The other region is called the *exterior* of the angle.

ANGLE BISECTOR: If C is in the interior of $\angle AOB$, _____

_____

_____

_____

When we say $m\angle AOC = m\angle COB$, we mean that the angle measures are equal.

Geometry Assumptions

In working with lines and angles, we again make specific assumptions that need to be identified. For example, in the definition of interior of an angle above, we assumed that an angle separated the plane into two disjoint sets. This follows from the assumption: *Given a line, the points of the plane that do not lie on the line form two sets called half-planes, such that (1) each of the sets is convex, and (2) if P is a point in one of the sets, and Q is a point in the other, then the segment PQ intersects the line.*

From this assumption, another obvious fact follows about a segment that intersects the sides of an angle: Given an $\angle AOB$, then for any point C in the interior of $\angle AOB$, the ray OC always intersects the segment AB.

S.12 Lesson 3: Copy and Bisect an Angle

This work is derived from Eureka Math ™ and licensed by Great Minds. ©2015 Great Minds. eureka-math.org
GEO-M1-SE-B1-1.3.0-05.2015

A STORY OF FUNCTIONS

In this lesson, we move from working with line segments to working with angles, specifically with bisecting angles. Before we do this, we need to clarify our assumptions about measuring angles. These assumptions are based upon what we know about a protractor that measures up to 180° angles:

1. To every ∠AOB there corresponds a quantity $m\angle AOB$ called the degree or measure of the angle so that $0° < m\angle AOB < 180°$.

This number, of course, can be thought of as the angle measurement (in degrees) of the interior part of the angle, which is what we read off of a protractor when measuring an angle. In particular, we have also seen that we can use protractors to "add angles":

2. If C is a point in the interior of ∠AOB, then $m\angle AOC + m\angle COB = m\angle AOB$.

Two angles ∠BAC and ∠CAD form a *linear pair* if $\overrightarrow{AB}$ and $\overrightarrow{AD}$ are opposite rays on a line, and $\overrightarrow{AC}$ is any other ray. In earlier grades, we abbreviated this situation and the fact that the measures of the angles on a line add up to 180° as, "∠'s on a line." Now, we state it formally as one of our assumptions:

3. If two angles ∠BAC and ∠CAD form a linear pair, then they are supplementary (i.e., $m\angle BAC + m\angle CAD = 180°$).

Protractors also help us to draw angles of a specified measure:

4. Let $\overrightarrow{OB}$ be a ray on the edge of the half-plane H. For every r such that $0° < r° < 180°$, there is exactly one ray OA with A in H such that $m\angle AOB = r°$.

Mathematical Modeling Exercise 1: Investigate How to Bisect an Angle

You need a compass and a straightedge.

Joey and his brother, Jimmy, are working on making a picture frame as a birthday gift for their mother. Although they have the wooden pieces for the frame, they need to find the angle bisector to accurately fit the edges of the pieces together. Using your compass and straightedge, show how the boys bisected the corner angles of the wooden pieces below to create the finished frame on the right.

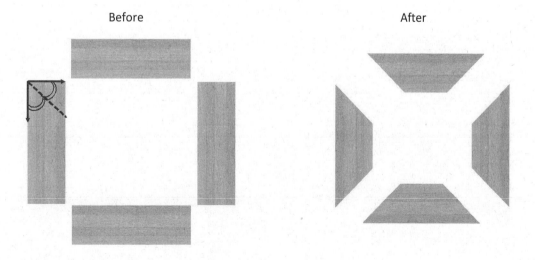

Before   After

Lesson 3:   Copy and Bisect an Angle

Consider how the use of circles aids the construction of an angle bisector. Be sure to label the construction as it progresses and to include the labels in your steps. Experiment with the angles below to determine the correct steps for the construction.

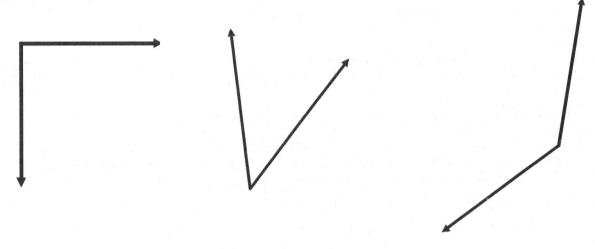

What steps did you take to bisect an angle? List the steps below:

_____

_____

_____

_____

_____

_____

_____

Lesson 3: Copy and Bisect an Angle

Mathematical Modeling Exercise 2: Investigate How to Copy an Angle

You will need a compass and a straightedge.

You and your partner will be provided with a list of steps (in random order) needed to copy an angle using a compass and straightedge. Your task is to place the steps in the correct order, then follow the steps to copy the angle below.

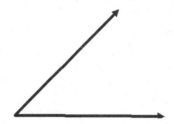

Steps needed (in correct order):

1. _____

2. _____

3. _____

4. _____

5. _____

6. _____

7. _____

8. _____

9. _____

Lesson 3: Copy and Bisect an Angle

Relevant Vocabulary

MIDPOINT: A point B is called a *midpoint* of $\overline{AC}$ if B is between A and C, and $AB = BC$.

DEGREE: Subdivide the length around a circle into 360 arcs of equal length. A central angle for any of these arcs is called a one-degree angle and is said to have angle measure 1 degree. An angle that turns through n one-degree angles is said to have an angle measure of n degrees.

ZERO AND STRAIGHT ANGLE: A *zero angle* is just a ray and measures 0°. A *straight angle* is a line and measures 180° (the ° is a symbol for *degree*).

Problem Set

Bisect each angle below.

1.
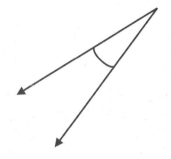

2.
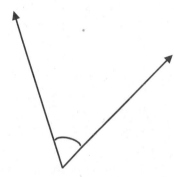

3.
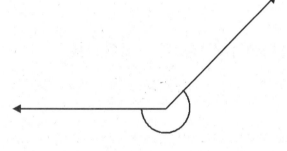

4.
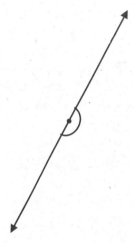

Copy the angle below.

5.
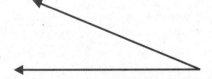

This page intentionally left blank

Lesson 4: Construct a Perpendicular Bisector

Classwork

Opening Exercise

Choose **one** method below to check your Problem Set:

- Trace your copied angles and bisectors onto patty paper; then, fold the paper along the bisector you constructed. Did one ray exactly overlap the other?
- Work with your partner. Hold one partner's work over another's. Did your angles and bisectors coincide perfectly?

Use the following rubric to evaluate your Problem Set:

Needs Improvement	Satisfactory	Excellent
Few construction arcs visible	Some construction arcs visible	Construction arcs visible and appropriate
Few vertices or relevant intersections labeled	Most vertices and relevant intersections labeled	All vertices and relevant intersections labeled
Lines drawn without straightedge or not drawn correctly	Most lines neatly drawn with straightedge	Lines neatly drawn with straightedge
Fewer than 3 angle bisectors constructed correctly	3 of the 4 angle bisectors constructed correctly	Angle bisector constructed correctly

Discussion

In Lesson 3, we studied how to construct an angle bisector. We know we can verify the construction by folding an angle along the bisector. A correct construction means that one half of the original angle coincides exactly with the other half so that each point of one ray of the angle maps onto a corresponding point on the other ray of the angle.

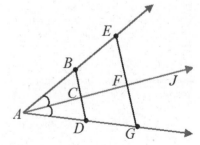

We now extend this observation. Imagine a segment that joins any pair of points that map onto each other when the original angle is folded along the bisector. The figure to the right illustrates two such segments.

Let us examine one of the two segments, $\overline{EG}$. When the angle is folded along $\overrightarrow{AJ}$, E coincides with G. In fact, folding the angle demonstrates that E is the same distance from F as G is from F; $EF = FG$. The point that separates these equal halves of $\overline{EG}$ is F, which is, in fact, the midpoint of the segment and lies on the bisector $\overrightarrow{AJ}$. We can make this case for any segment that falls under the conditions above.

By using geometry facts we acquired in earlier school years, we can also show that the angles formed by the segment and the angle bisector are right angles. Again, by folding, we can show that $\angle EFJ$ and $\angle GFJ$ coincide and must have the same measure. The two angles also lie on a straight line, which means they sum to $180°$. Since they are equal in measure and sum to $180°$, they each have a measure of $90°$.

These arguments lead to a remark about symmetry with respect to a line and the definition of a perpendicular bisector. Two points are symmetric with respect to a line l if and only if l is the perpendicular bisector of the segment that joins the two points. A perpendicular bisector of a segment passes through the _____ of the segment and forms _____ with the segment.

We now investigate how to construct a perpendicular bisector of a line segment using a compass and a straightedge. Using what you know about the construction of an angle bisector, experiment with your construction tools and the following line segment to establish the steps that determine this construction.

$$\overline{A B}$$

Precisely describe the steps you took to bisect the segment.

_____

_____

_____

_____

_____

Now that you are familiar with the construction of a perpendicular bisector, we must make one last observation. Using your compass, string, or patty paper, examine the following pairs of segments:

i. $\overline{AC}, \overline{BC}$
ii. $\overline{AD}, \overline{BD}$
iii. $\overline{AE}, \overline{BE}$

Based on your findings, fill in the observation below.

Observation:

Any point on the perpendicular bisector of a line segment is _____ from the endpoints of the line segment.

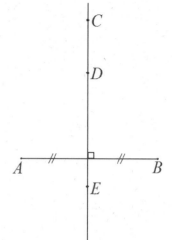

Mathematical Modeling Exercise

You know how to construct the perpendicular bisector of a segment. Now, investigate how to construct a perpendicular to a line ℓ from a point A not on ℓ. Think about how you have used circles in constructions so far and *why* the perpendicular bisector construction works the way it does. The first step of the instructions has been provided for you. Discover the construction, and write the remaining steps.

Step 1. Draw circle A: center A with radius so that circle A intersects line ℓ in two points.

_____

_____

_____

_____

_____

_____

Relevant Vocabulary

RIGHT ANGLE: An angle is called a *right angle* if its measure is 90°.

PERPENDICULAR: Two lines are *perpendicular* if they intersect in one point and if any of the angles formed by the intersection of the lines is a 90° (right) angle. Two segments or rays are perpendicular if the lines containing them are perpendicular lines.

EQUIDISTANT: A point A is said to be *equidistant* from two different points B and C if $AB = AC$. A point A is said to be *equidistant* from a point B and a line l if the distance between A and l is equal to AB.

Problem Set

1. During this lesson, you constructed a perpendicular line to a line ℓ from a point A not on ℓ. We are going to use that construction to construct parallel lines:

   To construct parallel lines ℓ_1 and ℓ_2:

   i. Construct a perpendicular line ℓ_3 to a line ℓ_1 from a point A not on ℓ_1.

   ii. Construct a perpendicular line ℓ_2 to ℓ_3 through point A. *Hint:* Consider using the steps behind Problem 4 in the Lesson 3 Problem Set to accomplish this.

   $A.$

   ℓ_1 ────────────────────────────────

Lesson 4: Construct a Perpendicular Bisector

2. Construct the perpendicular bisectors of $\overline{AB}$, $\overline{BC}$, and $\overline{CA}$ on the triangle below. What do you notice about the segments you have constructed?

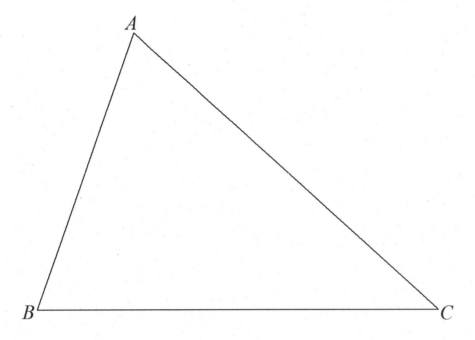

3. Two homes are built on a plot of land. Both homeowners have dogs and are interested in putting up as much fencing as possible between their homes on the land but in a way that keeps the fence equidistant from each home. Use your construction tools to determine where the fence should go on the plot of land. How must the fencing be altered with the addition of a third home?

This page intentionally left blank

Lesson 5: Points of Concurrencies

Classwork

Opening Exercise

You need a makeshift compass made from string and pencil.

Use these materials to construct the perpendicular bisectors of the three sides of the triangle below (like you did with Lesson 4, Problem Set 2).

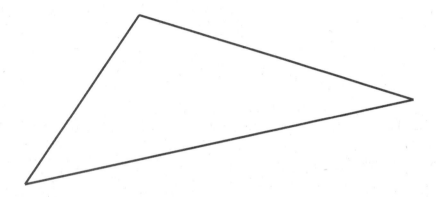

How did using this tool differ from using a compass and straightedge? Compare your construction with that of your partner. Did you obtain the same results?

_____

_____

Exploratory Challenge

When three or more lines intersect in a single point, they are _____, and the point of intersection is the _____.

You saw an example of a point of concurrency in yesterday's Problem Set (and in the Opening Exercise today) when all three perpendicular bisectors passed through a common point.

The point of concurrency of the three perpendicular bisectors is the _____.

The circumcenter of △ ABC is shown below as point P.

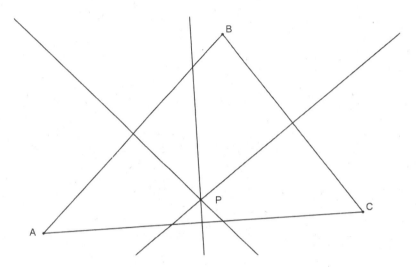

The questions that arise here are WHY are the three perpendicular bisectors concurrent? And WILL these bisectors be concurrent in all triangles? Recall that all points on the perpendicular bisector are equidistant from the endpoints of the segment, which means the following:

a. P is equidistant from A and B since it lies on the _____ of $\overline{AB}$.

b. P is also _____ from B and C since it lies on the perpendicular bisector of $\overline{BC}$.

c. Therefore, P must also be equidistant from A and C.

Hence, $AP = BP = CP$, which suggests that P is the point of _____ of all three perpendicular bisectors.

You have also worked with angle bisectors. The construction of the three angle bisectors of a triangle also results in a point of concurrency, which we call the _____.

Use the triangle below to construct the angle bisectors of each angle in the triangle to locate the triangle's incenter.

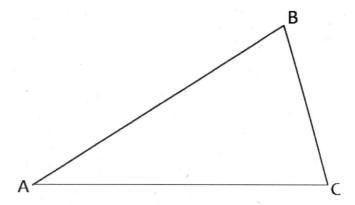

d. State precisely the steps in your construction above.

___

___

___

e. Earlier in this lesson, we explained why the perpendicular bisectors of the sides of a triangle are always concurrent. Using similar reasoning, explain clearly why the angle bisectors are always concurrent at the incenter of a triangle.

___

___

___

___

___

___

f. Observe the constructions below. Point A is the _____ of $\triangle JKL$. (Notice that it can fall outside of the triangle). Point B is the _____ of $\triangle RST$. The circumcenter of a triangle is the center of the circle that circumscribes that triangle. The incenter of the triangle is the center of the circle that is inscribed in that triangle.

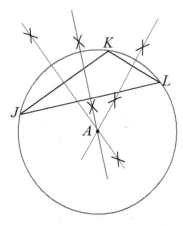
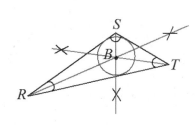

On a separate piece of paper, draw two triangles of your own below and demonstrate how the circumcenter and incenter have these special relationships.

g. How can you use what you have learned in Exercise 3 to find the center of a circle if the center is not shown?

_____

_____

_____

_____

Lesson 5: Points of Concurrencies

Problem Set

1. Given line segment AB, using a compass and straightedge, construct the set of points that are equidistant from A and B.

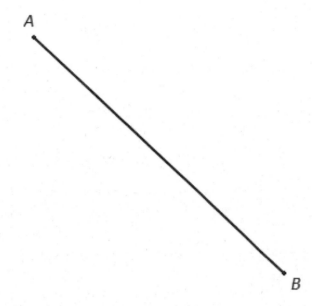

   What figure did you end up constructing? Explain.

2. For each of the following, construct a line perpendicular to segment AB that goes through point P.

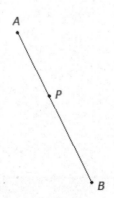 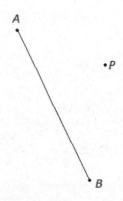

Lesson 5: Points of Concurrencies

3. Using a compass and straightedge, construct the angle bisector of ∠ABC shown below. What is true about every point that lies on the ray you created?

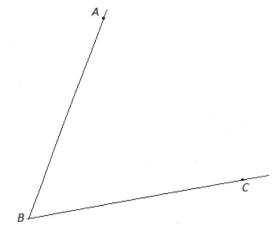

Lesson 6: Solve for Unknown Angles—Angles and Lines at a Point

Classwork

Opening Exercise

Determine the measure of the missing angle in each diagram.

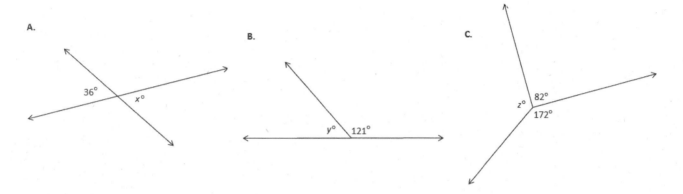

What facts about angles did you use?

Discussion

Two angles ∠AOC and ∠COB, with a common side $\overrightarrow{OC}$, are _____ if C belongs to the interior of ∠AOB. The sum of angles on a straight line is 180°, and two such angles are called a *linear pair*. Two angles are called *supplementary* if the sum of their measures is _____; two angles are called *complementary* if the sum of their measures is _____. Describing angles as supplementary or complementary refers only to the measures of their angles. The positions of the angles or whether the pair of angles is adjacent to each other is not part of the definition.

A STORY OF FUNCTIONS Lesson 6 M1
GEOMETRY

In the figure, line segment AD is drawn.
Find $m\angle DCE$.

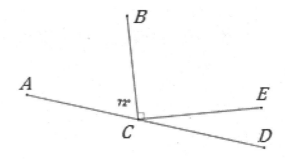

The total measure of adjacent angles around a point is _____.
Find the measure of $\angle HKI$.

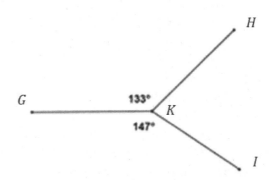

Vertical angles have _____ measure. Two angles are vertical if their sides form opposite rays.
Find $m\angle TRV$.

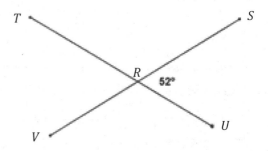

S.34 Lesson 6: Solve for Unknown Angles—Angles and Lines at a Point

A STORY OF FUNCTIONS   Lesson 6   M1

GEOMETRY

Example

Find the measures of each labeled angle. Give a reason for your solution.

 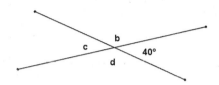 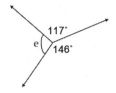

Angle	Angle Measure	Reason
∠a		
∠b		
∠c		
∠d		
∠e		

Exercises

In the figures below, $\overline{AB}$, $\overline{CD}$, and $\overline{EF}$ are straight line segments. Find the measure of each marked angle, or find the unknown numbers labeled by the variables in the diagrams. Give reasons for your calculations. Show all the steps to your solutions.

1.
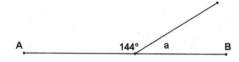

$m\angle a =$ _____

2.
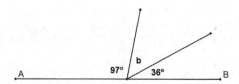

$m\angle b =$ _____

Lesson 6:   Solve for Unknown Angles—Angles and Lines at a Point   S.35

3.

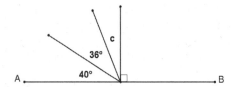

$m\angle c =$ _____

4.

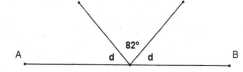

$m\angle d =$ _____

5.

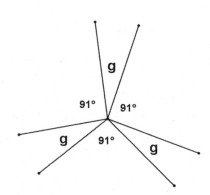

$m\angle g =$ _____

For Exercises 6–12, find the values of x and y. Show all work.

6.

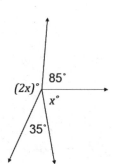

$x =$ _____

Lesson 6: Solve for Unknown Angles—Angles and Lines at a Point

7.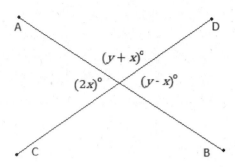

x = _____   y = _____

8.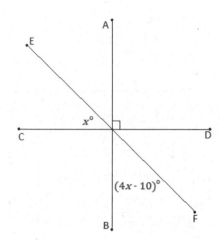

x = _____

9.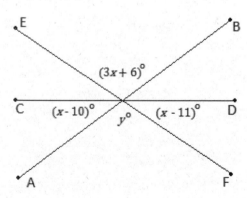

x = _____   y = _____

Lesson 6: Solve for Unknown Angles—Angles and Lines at a Point

10.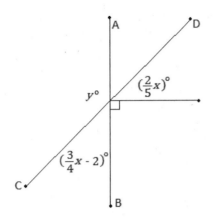

x = _____   y = _____

11.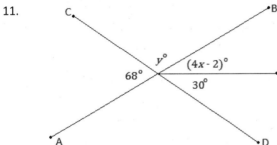

x = _____   y = _____

12.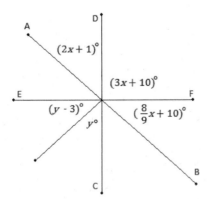

x = _____   y = _____

Relevant Vocabulary

STRAIGHT ANGLE: If two rays with the same vertex are distinct and collinear, then the rays form a line called a *straight angle*.

VERTICAL ANGLES: Two angles are *vertical angles* (or vertically opposite angles) if their sides form two pairs of opposite rays.

A STORY OF FUNCTIONS　　　　　　　　　　　　　　　　　　　　　　　　　Lesson 6　M1

GEOMETRY

Problem Set

In the figures below, $\overline{AB}$ and $\overline{CD}$ are straight line segments. Find the value of x and/or y in each diagram below. Show all the steps to your solutions, and give reasons for your calculations.

1.

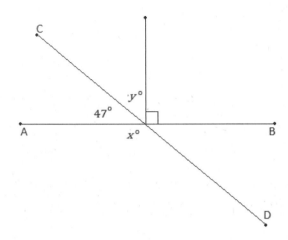

$x = $ _____

$y = $ _____

2.

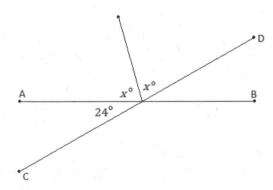

$x = $ _____

3.

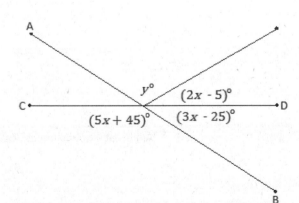

$x = $ _____

$y = $ _____

Lesson 6:　Solve for Unknown Angles—Angles and Lines at a Point

A STORY OF FUNCTIONS

Lesson 6 M1

GEOMETRY

Key Facts and Discoveries from Earlier Grades

Facts (With Abbreviations Used in Grades 4–9)	Diagram/Example	How to State as a Reason in an Exercise or a Proof
Vertical angles are equal in measure. (vert. ∠s)	$a° = b°$	"Vertical angles are equal in measure."
If C is a point in the interior of $\angle AOB$, then $m\angle AOC + m\angle COB = m\angle AOB$. (∠s add)	$m\angle AOB = m\angle AOC + m\angle COB$	"Angle addition postulate"
Two angles that form a linear pair are supplementary. (∠s on a line)	$a° + b° = 180$	"Linear pairs form supplementary angles."
Given a sequence of n consecutive adjacent angles whose interiors are all disjoint such that the angle formed by the first $n-1$ angles and the last angle are a linear pair, then the sum of all of the angle measures is 180°. (∠s on a line)	$a° + b° + c° + d° = 180$	"Consecutive adjacent angles on a line sum to 180°."
The sum of the measures of all angles formed by three or more rays with the same vertex and whose interiors do not overlap is 360°. (∠s at a point)	$m\angle ABC + m\angle CBD + m\angle DBA = 360°$	"Angles at a point sum to 360°."

Lesson 6: Solve for Unknown Angles—Angles and Lines at a Point

Facts (With Abbreviations Used in Grades 4–9)	Diagram/Example	How to State as a Reason in an Exercise or a Proof
The sum of the 3 angle measures of any triangle is 180°. (∠ sum of △)	$m\angle A + m\angle B + m\angle C = 180°$	"The sum of the angle measures in a triangle is 180°."
When one angle of a triangle is a right angle, the sum of the measures of the other two angles is 90°. (∠ sum of rt. △)	$m\angle A = 90°; m\angle B + m\angle C = 90°$	"Acute angles in a right triangle sum to 90°."
The sum of each exterior angle of a triangle is the sum of the measures of the opposite interior angles, or the remote interior angles. (ext. ∠ of △)	$m\angle BAC + m\angle ABC = m\angle BCD$	"The exterior angle of a triangle equals the sum of the two opposite interior angles."
Base angles of an isosceles triangle are equal in measure. (base ∠s of isos. △)		"Base angles of an isosceles triangle are equal in measure."
All angles in an equilateral triangle have equal measure. (equilat. △)		"All angles in an equilateral triangle have equal measure."

Facts (With Abbreviations Used in Grades 4–9)	Diagram/Example	How to State as a Reason in an Exercise or a Proof
If two parallel lines are intersected by a transversal, then corresponding angles are equal in measure. (corr. ∠s, $\overline{AB} \parallel \overline{CD}$)		"If parallel lines are cut by a transversal, then corresponding angles are equal in measure."
If two lines are intersected by a transversal such that a pair of corresponding angles are equal in measure, then the lines are parallel. (corr. ∠s converse)		"If two lines are cut by a transversal such that a pair of corresponding angles are equal in measure, then the lines are parallel."
If two parallel lines are intersected by a transversal, then interior angles on the same side of the transversal are supplementary. (int. ∠s, $\overline{AB} \parallel \overline{CD}$)		"If parallel lines are cut by a transversal, then interior angles on the same side are supplementary."
If two lines are intersected by a transversal such that a pair of interior angles on the same side of the transversal are supplementary, then the lines are parallel. (int. ∠s converse)		"If two lines are cut by a transversal such that a pair of interior angles on the same side are supplementary, then the lines are parallel."
If two parallel lines are intersected by a transversal, then alternate interior angles are equal in measure. (alt. ∠s, $\overline{AB} \parallel \overline{CD}$)		"If parallel lines are cut by a transversal, then alternate interior angles are equal in measure."
If two lines are intersected by a transversal such that a pair of alternate interior angles are equal in measure, then the lines are parallel. (alt. ∠s converse)		"If two lines are cut by a transversal such that a pair of alternate interior angles are equal in measure, then the lines are parallel."

Lesson 6: Solve for Unknown Angles—Angles and Lines at a Point

Lesson 7: Solve for Unknown Angles—Transversals

Classwork

Opening Exercise

Use the diagram at the right to determine x and y.
$\overleftrightarrow{AB}$ and $\overleftrightarrow{CD}$ are straight lines.

$x = $ _____

$y = $ _____

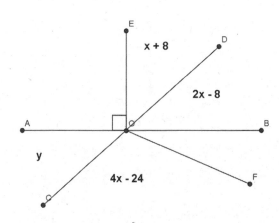

Name a pair of vertical angles:

_____

Find the measure of $\angle BOF$. Justify your calculation.

_____

_____

Discussion

Given line AB and line CD in a plane (see the diagram below), a third line EF is called a *transversal* if it intersects $\overleftrightarrow{AB}$ at a single point and intersects $\overleftrightarrow{CD}$ at a single but different point. Line AB and line CD are parallel if and only if the following types of angle pairs are congruent or supplementary.

- Corresponding angles are equal in measure.

  _____

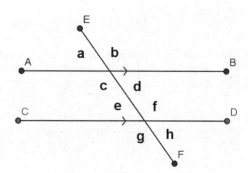

- Alternate interior angles are equal in measure.

  _____

- Same-side interior angles are supplementary.

  _____

Examples

1.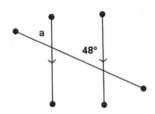

   $m\angle a =$ _____

2.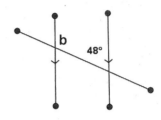

   $m\angle b =$ _____

3.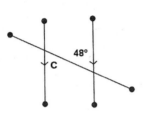

   $m\angle c =$ _____

4.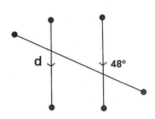

   $m\angle d =$ _____

5. An _____ is sometimes useful when solving for unknown angles.

   In this figure, we can use the auxiliary line to find the measures of $\angle e$ and $\angle f$ (how?) and then add the two measures together to find the measure of $\angle W$.

   What is the measure of $\angle W$?

   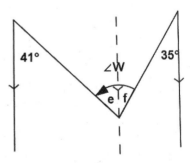

A STORY OF FUNCTIONS   Lesson 7   M1
                         GEOMETRY

Exercises 1–10

In each exercise below, find the unknown (labeled) angles. Give reasons for your solutions.

1.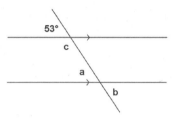

   $m\angle a = $ _____

   $m\angle b = $ _____

   $m\angle c = $ _____

2.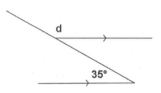

   $m\angle d = $ _____

3.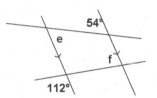

   $m\angle e = $ _____

   $m\angle f = $ _____

4.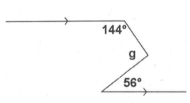

   $m\angle g = $ _____

5.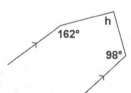

   $m\angle h = $ _____

6.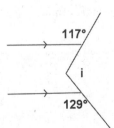

   $m\angle i = $ _____

Lesson 7:   Solve for Unknown Angles—Transversals

7.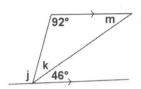

m∠j = _____

m∠k = _____

m∠m = _____

8.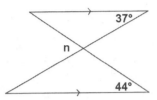

m∠n = _____

9.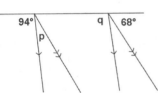

m∠p = _____

m∠q = _____

10.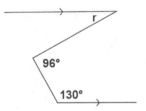

m∠r = _____

Relevant Vocabulary

ALTERNATE INTERIOR ANGLES: Let line t be a transversal to lines l and m such that t intersects l at point P and intersects m at point Q. Let R be a point on line l and S be a point on line m such that the points R and S lie in opposite half-planes of t. Then ∠RPQ and ∠PQS are called *alternate interior angles* of the transversal t with respect to line m and line l.

CORRESPONDING ANGLES: Let line t be a transversal to lines l and m. If ∠x and ∠y are alternate interior angles and ∠y and ∠z are vertical angles, then ∠x and ∠z are *corresponding angles*.

A STORY OF FUNCTIONS Lesson 7 M1
GEOMETRY

Problem Set

Find the unknown (labeled) angles. Give reasons for your solutions.

1.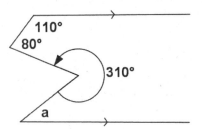

   $m\angle a =$ _____

2.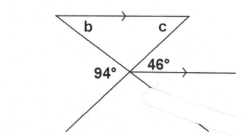

   $m\angle b =$ _____

   $m\angle c =$ _____

3.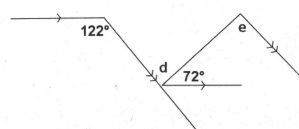

   $m\angle d =$ _____

   $m\angle e =$ _____

4.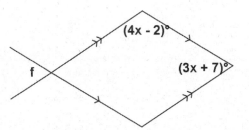

   $m\angle f =$ _____

Lesson 7: Solve for Unknown Angles—Transversals

This page intentionally left blank

Lesson 8: Solve for Unknown Angles—Angles in a Triangle

Classwork

Opening Exercise

Find the measure of angle x in the figure to the right. Explain your calculations. (Hint: Draw an auxiliary line segment.)

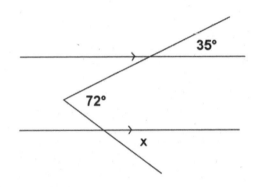

Discussion

The sum of the 3 angle measures of any triangle is _____.

INTERIOR OF A TRIANGLE: A point lies in the *interior of a triangle* if it lies in the interior of each of the angles of the triangle.

In any triangle, the measure of the exterior angle is equal to the sum of the measures of the _____ angles.

These are sometimes also known as _____ angles.

Base angles of an _____ triangle are equal in measure.

Each angle of an _____ triangle has a measure equal to 60°.

Relevant Vocabulary

ISOSCELES TRIANGLE: An *isosceles triangle* is a triangle with at least two sides of equal length.

ANGLES OF A TRIANGLE: Every triangle $\triangle ABC$ determines three angles, namely, $\angle BAC$, $\angle ABC$, and $\angle ACB$. These are called the *angles of* $\triangle ABC$.

EXTERIOR ANGLE OF A TRIANGLE: Let $\angle ABC$ be an interior angle of a triangle $\triangle ABC$, and let D be a point on $\overleftrightarrow{AB}$ such that B is between A and D. Then $\angle CBD$ is an *exterior angle of the triangle* $\triangle ABC$.

A STORY OF FUNCTIONS   Lesson 8 M1
                                                              GEOMETRY

Exercises 1–11

1. Find the measures of angles a and b in the figure to the right. Justify your results.

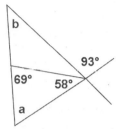

In each figure, determine the measures of the unknown (labeled) angles. Give reasons for your calculations.

2.

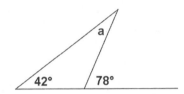

$m\angle a =$ _____

3.

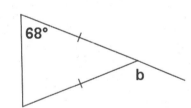

$m\angle b =$ _____

4.

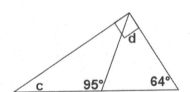

$m\angle c =$ _____

$m\angle d =$ _____

Lesson 8:   Solve for Unknown Angles—Angles in a Triangle

5.

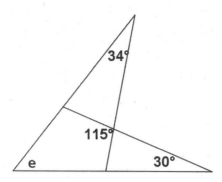

$m\angle e =$ _____

6.

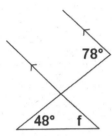

$m\angle f =$ _____

7.

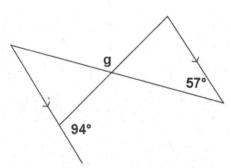

$m\angle g =$ _____

8.

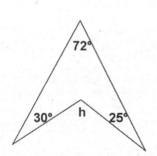

$m\angle h =$ _____

Lesson 8: Solve for Unknown Angles—Angles in a Triangle

9.

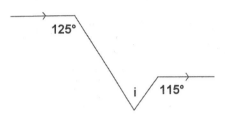

$m\angle i =$ _____

10.

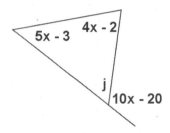

$m\angle j =$ _____

11.

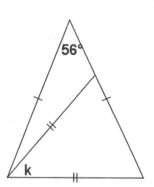

$m\angle k =$ _____

Problem Set

Find the unknown (labeled) angle in each figure. Justify your calculations.

1.

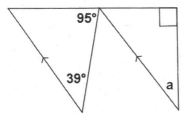

$m\angle a =$ _____

2.

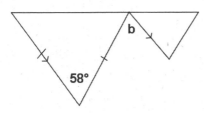

$m\angle b =$ _____

3.

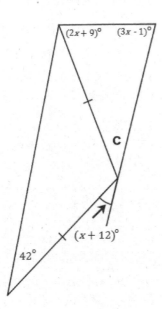

$m\angle c =$ _____

Lesson 8: Solve for Unknown Angles—Angles in a Triangle

This page intentionally left blank

Lesson 9: Unknown Angle Proofs—Writing Proofs

Classwork

Opening Exercise

One of the main goals in studying geometry is to develop your ability to reason critically, to draw valid conclusions based upon observations and proven facts. Master detectives do this sort of thing all the time. Take a look as Sherlock Holmes uses seemingly insignificant observations to draw amazing conclusions.

Could you follow Sherlock Holmes's reasoning as he described his thought process?

Discussion

In geometry, we follow a similar deductive thought process (much like Holmes uses) to prove geometric claims. Let's revisit an old friend—solving for unknown angles. Remember this one?

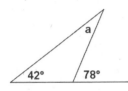

You needed to figure out the measure of a and used the "fact" that an exterior angle of a triangle equals the sum of the measures of the opposite interior angles. The measure of $\angle a$ must, therefore, be $36°$.

Suppose that we rearrange the diagram just a little bit.

Instead of using numbers, we use variables to represent angle measures.

Suppose further that we already know that the angles of a triangle sum to $180°$. Given the labeled diagram to the right, can we prove that $x + y = z$ (or, in other words, that the exterior angle of a triangle equals the sum of the measures of the opposite interior angles)?

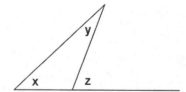

PROOF:

Label $\angle w$, as shown in the diagram.

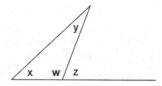

$m\angle x + m\angle y + m\angle w = 180°$	The sum of the angle measures in a triangle is $180°$.
$m\angle w + m\angle z = 180°$	Linear pairs form supplementary angles.
$m\angle x + m\angle y + m\angle w = m\angle w + m\angle z$	Substitution property of equality
$\therefore m\angle x + m\angle y = m\angle z$	Subtraction property of equality

Notice that each step in the proof was justified by a previously known or demonstrated fact. We end up with a newly proven fact (that an exterior angle of any triangle is the sum of the measures of the opposite interior angles of the triangle). This ability to identify the steps used to reach a conclusion based on known facts is *deductive reasoning* (i.e., the same type of reasoning that Sherlock Holmes used to accurately describe the doctor's attacker in the video clip).

Exercises 1–6

1. You know that angles on a line sum to $180°$.

   Prove that vertical angles are equal in measure.

   Make a plan:

   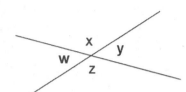

   - What do you know about $\angle w$ and $\angle x$? $\angle y$ and $\angle x$?

   - What conclusion can you draw based on both pieces of knowledge?

   - Write out your proof:

2. Given the diagram to the right, prove that $m\angle w + m\angle x + m\angle z = 180°$.
   (Make a plan first. What do you know about $\angle x$, $\angle y$, and $\angle z$?)

   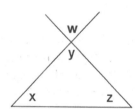

   Given the diagram to the right, prove that $m\angle w = m\angle y + m\angle z$.

   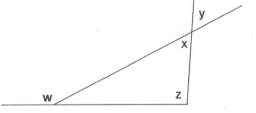

3. In the diagram to the right, prove that $m\angle y + m\angle z = m\angle w + m\angle x$.
   (You need to write a label in the diagram that is not labeled yet for this proof.)

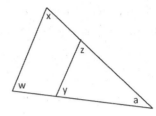

4. In the figure to the right, $\overline{AB} \parallel \overline{CD}$ and $\overline{BC} \parallel \overline{DE}$.
   Prove that $m\angle ABC = m\angle CDE$.

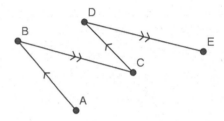

5. In the figure to the right, prove that the sum of the angles marked by arrows is 900°.
   (You need to write several labels in the diagram for this proof.)

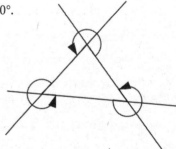

6. In the figure to the right, prove that $\overline{DC} \perp \overline{EF}$. Draw in label Z.

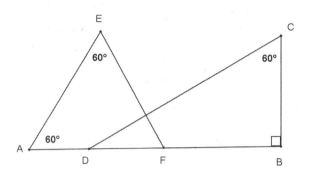

Problem Set

1. In the figure to the right, prove that $m \parallel n$.

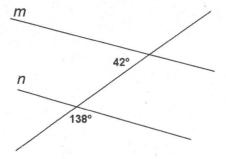

2. In the diagram to the right, prove that the sum of the angles marked by arrows is 360°.

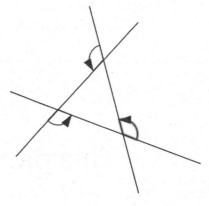

3. In the diagram to the right, prove that $m\angle a + m\angle d - m\angle b = 180°$.

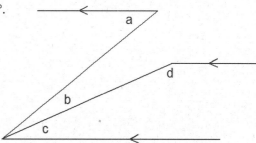

Lesson 9:   Unknown Angle Proofs—Writing Proofs

This page intentionally left blank

Basic Properties Reference Chart

Property	Meaning	Geometry Example
Reflexive Property	A quantity is equal to itself.	$AB = AB$
Transitive Property	If two quantities are equal to the same quantity, then they are equal to each other.	If $AB = BC$ and $BC = EF$, then $AB = EF$.
Symmetric Property	If a quantity is equal to a second quantity, then the second quantity is equal to the first.	If $OA = AB$, then $AB = OA$.
Addition Property of Equality	If equal quantities are added to equal quantities, then the sums are equal.	If $AB = DF$ and $BC = CD$, then $AB + BC = DF + CD$.
Subtraction Property of Equality	If equal quantities are subtracted from equal quantities, the differences are equal.	If $AB + BC = CD + DE$ and $BC = DE$, then $AB = CD$.
Multiplication Property of Equality	If equal quantities are multiplied by equal quantities, then the products are equal.	If $m\angle ABC = m\angle XYZ$, then $2(m\angle ABC) = 2(m\angle XYZ)$.
Division Property of Equality	If equal quantities are divided by equal quantities, then the quotients are equal.	If $AB = XY$, then $\dfrac{AB}{2} = \dfrac{XY}{2}$.
Substitution Property of Equality	A quantity may be substituted for its equal.	If $DE + CD = CE$ and $CD = AB$, then $DE + AB = CE$.
Partition Property (includes *Angle Addition Postulate*, *Segments Add*, *Betweenness of Points*, etc.)	A whole is equal to the sum of its parts.	If point C is on $\overline{AB}$, then $AC + CB = AB$.

This page intentionally left blank

Lesson 10: Unknown Angle Proofs—Proofs with Constructions

Classwork

Opening Exercise

In the figure on the right, $\overline{AB} \parallel \overline{DE}$ and $\overline{BC} \parallel \overline{EF}$. Prove that $b = e$.
(Hint: Extend $\overline{BC}$ and $\overline{ED}$.)

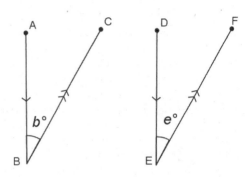

PROOF:

In the previous lesson, you used deductive reasoning with labeled diagrams to prove specific conjectures. What is different about the proof above?

Drawing or extending segments, lines, or rays (referred to as *auxiliary lines*) is frequently useful in demonstrating steps in the deductive reasoning process. Once $\overline{BC}$ and $\overline{ED}$ were extended, it was relatively simple to prove the two angles congruent based on our knowledge of alternate interior angles. Sometimes there are several possible extensions or additional lines that would work equally well.

For example, in this diagram, there are at least two possibilities for auxiliary lines. Can you spot them both?

Given: $\overline{AB} \parallel \overline{CD}$.
Prove: $z = x + y$.

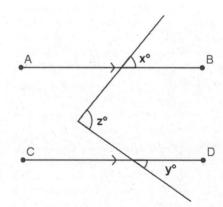

Discussion

Here is one possibility:

Given: $\overline{AB} \parallel \overline{CD}$.
Prove: $z = x + y$.

Extend the transversal as shown by the dotted line in the diagram. Label angle measures v and w, as shown.

What do you know about v and x?

About w and y? How does this help you?

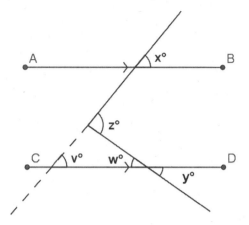

Write a proof using the auxiliary segment drawn in the diagram to the right.

Another possibility appears here:

Given: $\overline{AB} \parallel \overline{CD}$.
Prove: $z = x + y$.

Draw a segment parallel to $\overline{AB}$ through the vertex of the angle measuring z degrees. This divides the angle into two parts as shown.

What do you know about v and x?

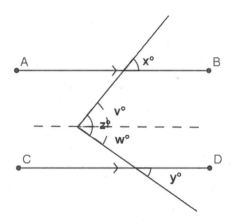

About w and y? How does this help you?

Write a proof using the auxiliary segment drawn in this diagram. Notice how this proof differs from the one above.

Examples

1. In the figure to the right, $\overline{AB} \parallel \overline{CD}$ and $\overline{BC} \parallel \overline{DE}$.
   Prove that $m\angle ABC = m\angle CDE$.
   (Is an auxiliary segment necessary?)

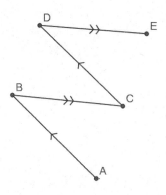

2. In the figure to the right, $\overline{AB} \parallel \overline{CD}$ and $\overline{BC} \parallel \overline{DE}$.
   Prove that $b + d = 180$.

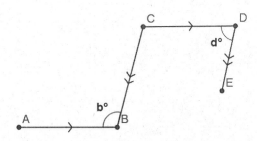

Lesson 10: Unknown Angle Proofs—Proofs with Constructions

3. In the figure to the right, prove that $d = a + b + c$.

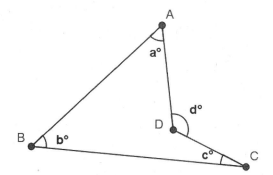

Problem Set

1. In the figure to the right, $\overline{AB} \parallel \overline{DE}$ and $\overline{BC} \parallel \overline{EF}$.
   Prove that $m\angle ABC = m\angle DEF$.

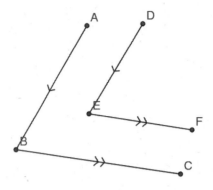

2. In the figure to the right, $\overline{AB} \parallel \overline{CD}$.
   Prove that $m\angle AEC = a° + c°$.

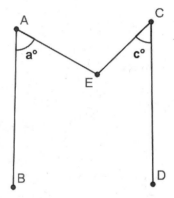

This page intentionally left blank

Lesson 11: Unknown Angle Proofs—Proofs of Known Facts

Classwork

Opening Exercise

A *proof* of a mathematical statement is a detailed explanation of how that statement follows logically from other statements already accepted as true.

A *theorem* is a mathematical statement with a proof.

Discussion

Once a theorem has been proved, it can be added to our list of known facts and used in proofs of other theorems. For example, in Lesson 9, we proved that *vertical angles are of equal measure*, and we know (from earlier grades and by paper cutting and folding) that *if a transversal intersects two parallel lines, alternate interior angles are of equal measure*. How do these facts help us prove that corresponding angles are equal in measure?

In the diagram to the right, if you are given that $\overline{AB} \parallel \overline{CD}$, how can you use your knowledge of how vertical angles and alternate interior angles are equal in measure to prove that $x = w$?

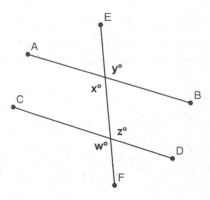

You now have available the following facts:

- Vertical angles are equal in measure.
- Alternate interior angles are equal in measure.
- Corresponding angles are equal in measure.

Use any or all of these facts to prove that *interior angles on the same side of the transversal are supplementary*. Add any necessary labels to the diagram below, and then write out a proof including given facts and a statement of what needs to be proved.

Given: $\overline{AB} \parallel \overline{CD}$, transversal $\overline{EF}$
Prove: $m\angle BGH + m\angle DHG = 180°$

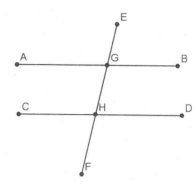

Now that you have proven this, you may add this theorem to your available facts.

- Interior angles on the same side of the transversal that intersects parallel lines sum to $180°$.

Use any of these four facts to prove that the three angles of a triangle sum to $180°$. For this proof, you will need to draw an auxiliary line, parallel to one of the triangle's sides and passing through the vertex opposite that side. Add any necessary labels, and write out your proof.

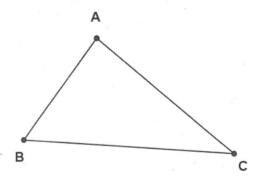

Let's review the theorems we have now proven:
- Vertical angles are equal in measure.
- A transversal intersects a pair of lines. The pair of lines is parallel if and only if:
  - Alternate interior angles are equal in measure.
  - Corresponding angles are equal in measure.
- Interior angles on the same side of the transversal add to 180°. The sum of the degree measures of the angles of a triangle is 180°.

Side Trip: Take a moment to take a look at one of those really famous Greek guys we hear so much about in geometry, Eratosthenes. Over 2,000 years ago, Eratosthenes used the geometry we have just been working with to find the circumference of Earth. He did not have cell towers, satellites, or any other advanced instruments available to scientists today. The only things Eratosthenes used were his eyes, his feet, and perhaps the ancient Greek equivalent to a protractor.

Watch this video to see how he did it, and try to spot the geometry we have been using throughout this lesson.

https://youtu.be/wnElDaV4esg

Example 1

Construct a proof designed to demonstrate the following:

If two lines are perpendicular to the same line, they are parallel to each other.

(a) Draw and label a diagram, (b) state the given facts and the conjecture to be proved, and (c) write out a clear statement of your reasoning to justify each step.

Discussion

Each of the three parallel line theorems has a converse (or reversing) theorem as follows:

Original	Converse
If two parallel lines are cut by a transversal, then alternate interior angles are equal in measure.	If two lines are cut by a transversal such that alternate interior angles are equal in measure, then the lines are parallel.
If two parallel lines are cut by a transversal, then corresponding angles are equal in measure.	If two lines are cut by a transversal such that corresponding angles are equal in measure, then the lines are parallel.
If two parallel lines are cut by a transversal, then interior angles on the same side of the transversal add to 180°.	If two lines are cut by a transversal such that interior angles on the same side of the transversal add to 180°, then the lines are parallel.

Notice the similarities between the statements in the first column and those in the second. Think about when you would need to use the statements in the second column, that is, the times when you are trying to prove two lines are parallel.

Example 2

In the figure to the right, $x = y$.
Prove that $\overline{AB} \parallel \overline{EF}$.

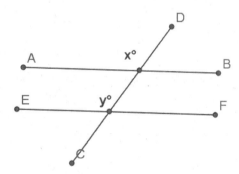

Lesson 11: Unknown Angle Proofs—Proofs of Known Facts

Problem Set

1. Given: ∠C and ∠D are supplementary and $m\angle B = m\angle D$
   Prove: $\overline{AB} \parallel \overline{CD}$

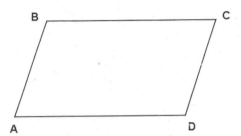

2. A theorem states that *in a plane, if a line is perpendicular to one of two parallel lines and intersects the other, then it is perpendicular to the other of the two parallel lines.*

   Prove this theorem. (a) Construct and label an appropriate figure, (b) state the given information and the theorem to be proven, and (c) list the necessary steps to demonstrate the proof.

This page intentionally left blank

Lesson 12: Transformations—The Next Level

Classwork

Opening Exercises

a. Find the measure of each lettered angle in the figure below.

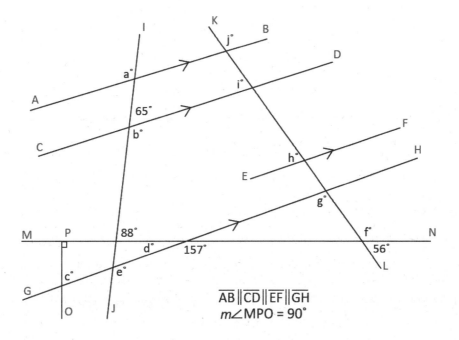

$a =$	$b =$	$c =$	$d =$	$e =$
$f =$	$g =$	$h =$	$i =$	$j =$

b. Given: $m\angle CDE = m\angle BAC$
Prove: $m\angle DEC = m\angle ABC$

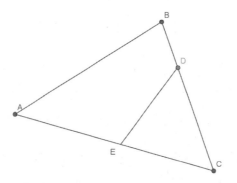

Mathematical Modeling Exercise

You will work with a partner on this exercise and are allowed a protractor, compass, and straightedge.

- **Partner A**: Use the card your teacher gives you. Without showing the card to your partner, describe to your partner how to draw the transformation indicated on the card. When you have finished, compare your partner's drawing with the transformed image on your card. Did you describe the motion correctly?

- **Partner B**: Your partner is going to describe a transformation to be performed on the figure on your card. Follow your partner's instructions and then compare the image of your transformation to the image on your partner's card.

Discussion

Explaining how to transform figures without the benefit of a coordinate plane can be difficult without some important vocabulary. Let's review.

The word *transformation* has a specific meaning in geometry. A transformation F of the plane is a function that assigns to each point P of the plane a unique point $F(P)$ in the plane. Transformations that preserve lengths of segments and measures of angles are called _____. A *dilation* is an example of a transformation that preserves _____ measures but not the lengths of segments. In this lesson, we work only with rigid transformations. We call a figure that is about to undergo a transformation the _____, while the figure that has undergone the transformation is called the _____.

A STORY OF FUNCTIONS   Lesson 12 M1

GEOMETRY

Rotation

Reflection

Translation

Using the figures above, identify specific information needed to perform the rigid motion shown.

For a rotation, we need to know:

_____

_____

For a reflection, we need to know:

_____

_____

For a translation, we need to know:

_____

_____

Lesson 12:   Transformations—The Next Level

Geometry Assumptions

We have now done some work with all three basic types of rigid motions (rotations, reflections, and translations). At this point, we need to state our assumptions as to the properties of basic rigid motions:

a. Any basic rigid motion preserves lines, rays, and segments. That is, for a basic rigid motion of the plane, the image of a line is a line, the image of a ray is a ray, and the image of a segment is a segment.

b. Any basic rigid motion preserves lengths of segments and measures of angles.

Relevant Vocabulary

BASIC RIGID MOTION: A *basic rigid motion* is a rotation, reflection, or translation of the plane.

Basic rigid motions are examples of transformations. Given a transformation, the image of a point A is the point the transformation maps A to in the plane.

DISTANCE-PRESERVING: A transformation is said to be *distance-preserving* if the distance between the images of two points is always equal to the distance between the pre-images of the two points.

ANGLE-PRESERVING: A transformation is said to be *angle-preserving* if (1) the image of any angle is again an angle and (2) for any given angle, the angle measure of the image of that angle is equal to the angle measure of the pre-image of that angle.

A STORY OF FUNCTIONS   Lesson 12   M1
GEOMETRY

Problem Set

An example of a rotation applied to a figure and its image are provided. Use this representation to answer the questions that follow. For each question, a pair of figures (pre-image and image) is given as well as the center of rotation. For each question, identify and draw the following:

  i. The circle that determines the rotation, using any point on the pre-image and its image.
  ii. An angle, created with three points of your choice, which demonstrates the angle of rotation.

Example of a Rotation:

Pre-image: (solid line)

Image: (dotted line)

Center of rotation: P

Angle of rotation: $\angle APA'$

1. Pre-image: (solid line)
   Image: (dotted line)
   Center of rotation: P

   Angle of rotation:_____

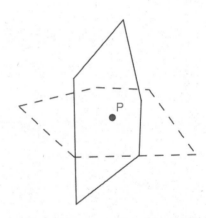

2. Pre-image: $\triangle ABC$
   Image: $\triangle A'B'C'$
   Center: D

   Angle of rotation:_____

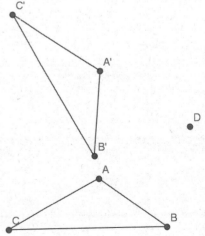

Lesson 12: Transformations—The Next Level

This page intentionally left blank

Lesson 13: Rotations

Classwork

Exploratory Challenge

You need a pair of scissors and a ruler.

Cut out the 75° angle on the right and use it as a guide to rotate the figure below 75° counterclockwise around the given center of rotation (Point P).

- Place the vertex of the 75° angle at point P.
- Line up one ray of the 75° angle with vertex A on the figure. Carefully measure the length from point P to vertex A.
- Measure that same distance along the other ray of the reference angle, and mark the location of your new point, A'.
- Repeat these steps for each vertex of the figure, labeling the new vertices as you find them.
- Connect the six segments that form the sides of your rotated image.

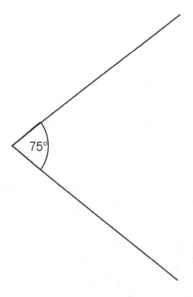

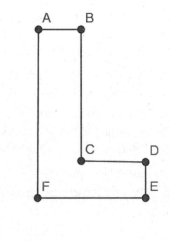

Discussion

In Grade 8, we spent time developing an understanding of what happens in the application of a rotation by participating in hands-on lessons. Now, we can define rotation precisely.

First, we need to talk about the direction of the rotation. If you stand up and spin in place, you can either spin to your left or spin to your right. This spinning to your left or right can be rephrased using what we know about analog clocks: spinning to your left is spinning in a counterclockwise direction, and spinning to your right is spinning in a clockwise direction. We need to have the same sort of notion for rotating figures in the plane. It turns out that there is a way to always choose a counterclockwise half-plane for any ray: The counterclockwise half-plane of $\overrightarrow{CP}$ is the half-plane of $\overrightarrow{CP}$ that lies to the left as you move along $\overrightarrow{CP}$ in the direction from C to P. (The clockwise half-plane is then the half-plane that lies to the right as you move along $\overrightarrow{CP}$ in the direction from C to P.) We use this idea to state the definition of rotation.

For $0° < \theta° < 180°$, the rotation of θ degrees around the center C is the transformation $R_{C,\theta}$ of the plane defined as follows:

1. For the center point C, $R_{C,\theta}(C) = C$, and
2. For any other point P, $R_{C,\theta}(P)$ is the point Q that lies in the counterclockwise half-plane of $\overrightarrow{CP}$, such that $CQ = CP$ and $m\angle PCQ = \theta°$.

A rotation of 0 degrees around the center C is the identity transformation (i.e., for all points A in the plane, it is the rotation defined by the equation $R_{C,0}(A) = A$).

A rotation of 180° around the center C is the composition of two rotations of 90° around the center C. It is also the transformation that maps every point P (other than C) to the other endpoint of the diameter of a circle with center C and radius CP.

- A rotation leaves the center point C fixed. $R_{C,\theta}(C) = C$ states exactly that. The rotation function R with center point C that moves everything else in the plane $\theta°$, leaves only the center point itself unmoved.
- Any other point P in the plane moves the exact same degree arc along the circle defined by the center of rotation and the angle $\theta°$.
- Then $R_{C,\theta}(P)$ is the point Q that lies in the counterclockwise half-plane of ray $\overrightarrow{CP}$ such that $CQ = CP$ and such that $m\angle PCQ = \theta°$. Visually, you can imagine rotating the point P in a counterclockwise arc around a circle with center C and radius CP to find the point Q.
- All positive angle measures θ assume a counterclockwise motion; if citing a clockwise rotation, the answer should be labeled with CW.

A composition of two rotations applied to a point is the image obtained by applying the second rotation to the image of the first rotation of the point. In mathematical notation, the image of a point A after *a composition of two rotations of 90° around the center C* can be described by the point $R_{C,90}\left(R_{C,90}(A)\right)$. The notation reads, "Apply $R_{C,90}$ to the point $R_{C,90}(A)$." So, we lose nothing by defining $R_{C,180}(A)$ to be that image. Then, $R_{C,180}(A) = R_{C,90}\left(R_{C,90}(A)\right)$ for all points A in the plane.

In fact, we can generalize this idea to define a rotation by any positive degree: For $\theta° > 180°$, a *rotation of $\theta°$ around the center C* is any composition of three or more rotations, such that each rotation is less than or equal to a 90° rotation and whose angle measures sum to $\theta°$. For example, a rotation of 240° is equal to the composition of three rotations by 80° about the same center, the composition of five rotations by 50°, 50°, 50°, 50°, and 40° about the same center, or the composition of 240 rotations by 1° about the same center.

Notice that we have been assuming that all rotations rotate in the counterclockwise direction. However, the inverse rotation (the rotation that *undoes* a given rotation) can be thought of as rotating in the clockwise direction. For example, rotate a point A by 30° around another point C to get the image $R_{C,30}(A)$. We can *undo* that rotation by rotating by 30° in the clockwise direction around the same center C. Fortunately, we have an easy way to describe a *rotation in the clockwise direction*. If all positive degree rotations are in the counterclockwise direction, then we can define a negative degree rotation as a rotation in the clockwise direction (using the clockwise half-plane instead of the counterclockwise half-plane). Thus, $R_{C,-30}$ is a 30° rotation in the clockwise direction around the center C. Since a composition of two rotations around the same center is just the sum of the degrees of each rotation, we see that

$$R_{C,-30}\left(R_{C,30}(A)\right) = R_{C,0}(A) = A,$$

for all points A in the plane. Thus, we have defined how to perform a rotation for any number of degrees—positive or negative.

As this is our first foray into close work with rigid motions, we emphasize an important fact about rotations. Rotations are one kind of rigid motion or transformation of the plane (a function that assigns to each point P of the plane a unique point $F(P)$) that preserves lengths of segments and measures of angles. Recall that Grade 8 investigations involved manipulatives that modeled rigid motions (e.g., transparencies) because you could actually *see* that a figure was not altered, as far as length or angle was concerned. It is important to hold onto this idea while studying all of the rigid motions.

Constructing rotations precisely can be challenging. Fortunately, computer software is readily available to help you create transformations easily. Geometry software (such as Geogebra) allows you to create plane figures and rotate them a given number of degrees around a specified center of rotation. The figures in the exercises were rotated using Geogebra. Determine the angle and direction of rotation that carries each pre-image onto its (dashed-line) image. Assume both angles of rotation are positive. The center of rotation for Exercise 1 is point D and for Figure 2 is point E.

Exercises 1–3

1.

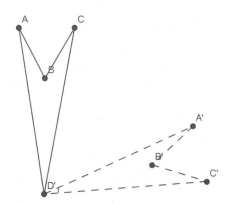

To determine the angle of rotation, you measure the angle formed by connecting corresponding vertices to the center point of rotation. In Exercise 1, measure $\angle AD'A'$. What happened to $\angle D$? Can you see that D is the center of rotation, therefore, mapping D' onto itself? Before leaving Exercise 1, try drawing $\angle BD'B'$. Do you get the same angle measure? What about $\angle CD'C'$?

Try finding the angle and direction of rotation for Exercise 2 on your own.

2.

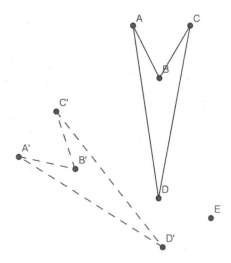

Did you draw $\angle DED'$ or $\angle CEC'$?

Now that you can find the angle of rotation, let's move on to finding the center of rotation. Follow the directions below to locate the center of rotation, taking the figure at the top right to its image at the bottom left.

3.

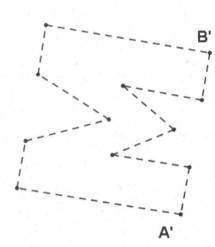

a. Draw a segment connecting points A and A'.

b. Using a compass and straightedge, find the perpendicular bisector of this segment.

c. Draw a segment connecting points B and B'.

d. Find the perpendicular bisector of this segment.

e. The point of intersection of the two perpendicular bisectors is the center of rotation. Label this point P.

Justify your construction by measuring $\angle APA'$ and $\angle BPB'$. Did you obtain the same measure?

Exercises 4–5

Find the centers of rotation and angles of rotation for Exercises 4 and 5.

4.

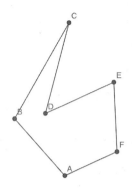
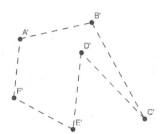

5.

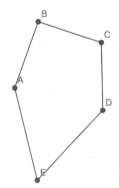
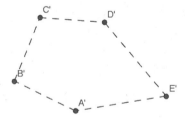

Lesson 13: Rotations

A STORY OF FUNCTIONS

Lesson 13   M1

GEOMETRY

> **Lesson Summary**
>
> A rotation carries segments onto segments of equal length.
>
> A rotation carries angles onto angles of equal measure.

Problem Set

1. Rotate triangle ABC 60° around point F using a compass and straightedge only.

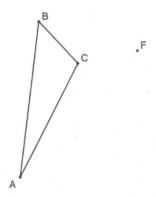

2. Rotate quadrilateral $ABCD$ 120° around point E using a straightedge and protractor.

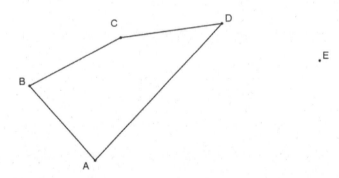

3. On your paper, construct a 45° angle using a compass and straightedge. Rotate the angle 180° around its vertex, again using only a compass and straightedge. What figure have you formed, and what are its angles called?

4. Draw a triangle with angles 90°, 60°, and 30° using only a compass and straightedge. Locate the midpoint of the longest side using your compass. Rotate the triangle 180° around the midpoint of the longest side. What figure have you formed?

Lesson 13:   Rotations

S.87

5. On your paper, construct an equilateral triangle. Locate the midpoint of one side using your compass. Rotate the triangle 180° around this midpoint. What figure have you formed?

6. Use either your own initials (typed using WordArt in Microsoft Word) or the initials provided below. If you create your own WordArt initials, copy, paste, and rotate to create a design similar to the one below. Find the center of rotation and the angle of rotation for your rotation design.

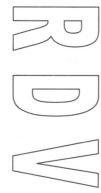

Lesson 14: Reflections

Classwork

Exploratory Challenge

Think back to Lesson 12 where you were asked to describe to your partner how to reflect a figure across a line. The greatest challenge in providing the description was using the precise vocabulary necessary for accurate results. Let's explore the language that yields the results we are looking for.

△ ABC is reflected across $\overline{DE}$ and maps onto △ $A'B'C'$.

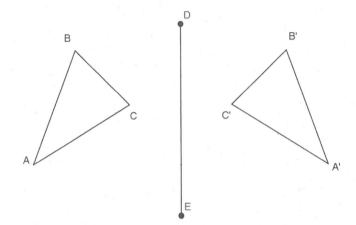

Use your compass and straightedge to construct the perpendicular bisector of each of the segments connecting A to A', B to B', and C to C'. What do you notice about these perpendicular bisectors?

Label the point at which $\overline{AA'}$ intersects $\overline{DE}$ as point O. What is true about AO and $A'O$? How do you know this is true?

Discussion

You just demonstrated that the line of reflection between a figure and its reflected image is also the perpendicular bisector of the segments connecting corresponding points on the figures.

In the Exploratory Challenge, you were given the pre-image, the image, and the line of reflection. For your next challenge, try finding the line of reflection provided a pre-image and image.

Example 1

Construct the segment that represents the line of reflection for quadrilateral $ABCD$ and its image $A'B'C'D'$.

What is true about each point on $ABCD$ and its corresponding point on $A'B'C'D'$ with respect to the line of reflection?

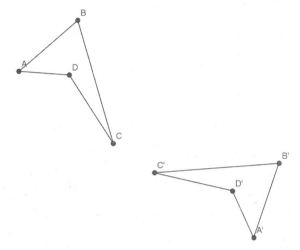

Notice one very important fact about reflections. Every point in the original figure is carried to a corresponding point on the image by the same rule—a reflection across a specific line. This brings us to a critical definition:

REFLECTION: For a line l in the plane, a *reflection across l* is the transformation r_l of the plane defined as follows:

1. For any point P on the line l, $r_l(P) = P$, and
2. For any point P not on l, $r_l(P)$ is the point Q so that l is the perpendicular bisector of the segment PQ.

If the line is specified using two points, as in $\overleftrightarrow{AB}$, then the reflection is often denoted by $r_{\overleftrightarrow{AB}}$. Just as we did in the last lesson, let's examine this definition more closely:

- *A transformation of the plane*—the entire plane is transformed; what was once on one side of the line of reflection is now on the opposite side;
- $r_l(P) = P$ *means that the points on line l are left fixed*—the only part of the entire plane that is left fixed is the line of reflection itself;
- $r_l(P)$ *is the point Q*—the transformation r_l maps the point P to the point Q;
- *The line of reflection l is the perpendicular bisector of the segment PQ*—to find Q, first construct the perpendicular line m to the line l that passes through the point P. Label the intersection of l and m as N. Then locate the point Q on m on the other side of l such that $PN = NQ$.

A STORY OF FUNCTIONS Lesson 14 M1
GEOMETRY

Examples 2–3

Construct the line of reflection across which each image below was reflected.

2.

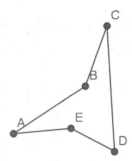

3.

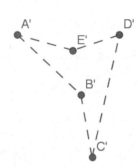

You have shown that a line of reflection is the perpendicular bisector of segments connecting corresponding points on a figure and its reflected image. You have also constructed a line of reflection between a figure and its reflected image. Now we need to explore methods for constructing the reflected image itself. The first few steps are provided for you in this next stage.

Lesson 14: Reflections

A STORY OF FUNCTIONS

Lesson 14 M1

GEOMETRY

Example 4

The task at hand is to construct the reflection of △ ABC over $\overline{DE}$. Follow the steps below to get started; then complete the construction on your own.

1. Construct circle A: center A, with radius such that the circle crosses $\overline{DE}$ at two points (labeled F and G).
2. Construct circle F: center F, radius FA and circle G: center G, radius GA. Label the (unlabeled) point of intersection between circles F and G as point A'. This is the reflection of vertex A across $\overline{DE}$.
3. Repeat steps 1 and 2 for vertices B and C to locate B' and C'.
4. Connect A', B', and C' to construct the reflected triangle.

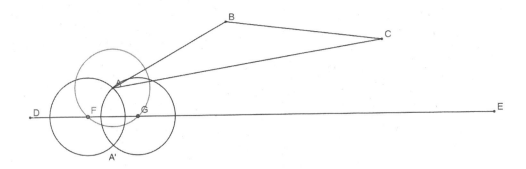

Things to consider:

When you found the line of reflection earlier, you did this by constructing perpendicular bisectors of segments joining two corresponding vertices. How does the reflection you constructed above relate to your earlier efforts at finding the line of reflection itself? Why did the construction above work?

Example 5

Now try a slightly more complex figure. Reflect $ABCD$ across $\overline{EF}$.

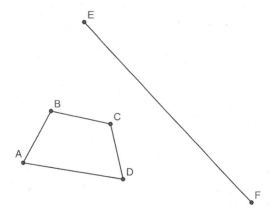

S.92   Lesson 14:   Reflections

EUREKA MATH

A STORY OF FUNCTIONS   Lesson 14   M1

GEOMETRY

> **Lesson Summary**
>
> A reflection carries segments onto segments of equal length.
>
> A reflection carries angles onto angles of equal measure.

Problem Set

Construct the line of reflection for each pair of figures below.

1.

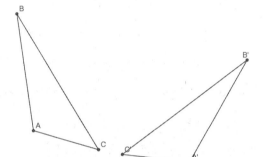

2.

3.

4. Reflect the given image across the line of reflection provided.

5. Draw a triangle ABC. Draw a line l through vertex C so that it intersects the triangle at more than just the vertex. Construct the reflection across l.

Lesson 14: Reflections

This page intentionally left blank

Lesson 15: Rotations, Reflections, and Symmetry

Classwork

Opening Exercise

The original triangle, labeled *A*, has been reflected across the first line, resulting in the image labeied *B*. Reflect the image across the second line.

Carlos looked at the image of the reflection across the second line and said, "That's not the image of triangle *A* after two reflections; that's the image of triangle *A* after a rotation!" Do you agree? Why or why not?

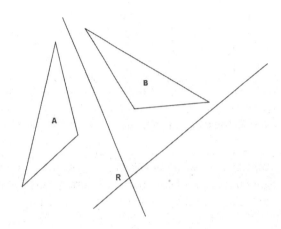

Discussion

When you reflect a figure across a line, the original figure and its image share a line of symmetry, which we have called the line of reflection. When you reflect a figure across a line and then reflect the image across a line that intersects the first line, your final image is a rotation of the original figure. The center of rotation is the point at which the two lines of reflection intersect. The angle of rotation is determined by connecting the center of rotation to a pair of corresponding vertices on the original figure and the final image. The figure above is a 210° rotation (or 150° clockwise rotation).

Exploratory Challenge

LINE OF SYMMETRY OF A FIGURE: This is an isosceles triangle. By definition, an isosceles triangle has at least two congruent sides. A *line of symmetry* of the triangle can be drawn from the top vertex to the midpoint of the base, decomposing the original triangle into two congruent right triangles. This line of symmetry can be thought of as a reflection across itself that takes the isosceles triangle to itself. Every point of the triangle on one side of the line of symmetry has a corresponding point on the triangle on the other side of the line of symmetry, given by reflecting the point across the line. In particular, the line of symmetry is equidistant from all corresponding pairs of points. Another way of thinking about line symmetry is that a figure has line symmetry if there exists a line (or lines) such that the image of the figure when reflected over the line is itself.

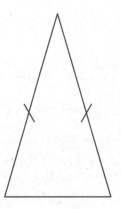

Does every figure have a line of symmetry?

Which of the following have multiple lines of symmetry?

Use your compass and straightedge to draw one line of symmetry on each figure above that has at least one line of symmetry. Then, sketch any remaining lines of symmetry that exist. What did you do to justify that the lines you constructed were, in fact, lines of symmetry? How can you be certain that you have found all lines of symmetry?

ROTATIONAL SYMMETRY OF A FIGURE: A nontrivial *rotational symmetry of a figure* is a rotation of the plane that maps the figure back to itself such that the rotation is greater than 0° but less than 360°. Three of the four polygons above have a nontrivial rotational symmetry. Can you identify the polygon that does not have such symmetry?

When we studied rotations two lessons ago, we located both a center of rotation and an angle of rotation.

Identify the center of rotation in the equilateral triangle ABC below, and label it D. Follow the directions in the paragraph below to locate the center precisely.

To identify the center of rotation in the equilateral triangle, the simplest method is finding the perpendicular bisector of at least two of the sides. The intersection of these two bisectors gives us the center of rotation. Hence, the center of rotation of an equilateral triangle is also the circumcenter of the triangle. In Lesson 5 of this module, you also located another special point of concurrency in triangles—the incenter. What do you notice about the incenter and circumcenter in the equilateral triangle?

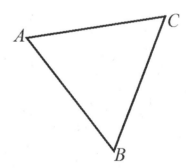

In any regular polygon, how do you determine the angle of rotation? Use the equilateral triangle above to determine the method for calculating the angle of rotation, and try it out on the rectangle, hexagon, and parallelogram above.

IDENTITY SYMMETRY: A symmetry of a figure is a basic rigid motion that maps the figure back onto itself. There is a special transformation that trivially maps any figure in the plane back to itself called the *identity transformation*. This transformation, like the function f defined on the real number line by the equation $f(x) = x$, maps each point in the plane back to the same point (in the same way that f maps 3 to 3, π to π, and so forth). It may seem strange to discuss the do-nothing *identity symmetry* (the symmetry of a figure under the identity transformation), but it is actually quite useful when listing all of the symmetries of a figure.

Let us look at an example to see why. The equilateral triangle ABC on the previous page has two nontrivial rotations about its circumcenter D, a rotation by $120°$ and a rotation by $240°$. Notice that performing two $120°$ rotations back-to-back is the same as performing one $240°$ rotation. We can write these two back-to-back rotations explicitly, as follows:

- First, rotate the triangle by $120°$ about D: $R_{D,120°}(\triangle ABC)$.
- Next, rotate the image of the first rotation by $120°$: $R_{D,120°}\big(R_{D,120°}(\triangle ABC)\big)$.

Rotating $\triangle ABC$ by $120°$ twice in a row is the same as rotating $\triangle ABC$ once by $120° + 120° = 240°$. Hence, rotating by $120°$ twice is equivalent to one rotation by $240°$:

$$R_{D,120°}\big(R_{D,120°}(\triangle ABC)\big) = R_{D,240°}(\triangle ABC).$$

In later lessons, we see that this can be written compactly as $R_{D,120°} \cdot R_{D,120°} = R_{D,240°}$. What if we rotated by $120°$ one more time? That is, what if we rotated $\triangle ABC$ by $120°$ three times in a row? That would be equivalent to rotating $\triangle ABC$ once by $120° + 120° + 120°$ or $360°$. But a rotation by $360°$ is equivalent to doing nothing (i.e., the identity transformation)! If we use I to denote the identity transformation ($I(P) = P$ for every point P in the plane), we can write this equivalency as follows:

$$R_{D,120°}\Big(R_{D,120°}\big(R_{D,120°}(\triangle ABC)\big)\Big) = I(\triangle ABC).$$

Continuing in this way, we see that rotating $\triangle ABC$ by $120°$ four times in a row is the same as rotating once by $120°$, rotating five times in a row is the same as $R_{D,240°}$, and so on. In fact, for a whole number n, rotating $\triangle ABC$ by $120° \, n$ times in a row is equivalent to performing one of the following three transformations:

$$\{R_{D,120°}, \quad R_{D,240°}, \quad I\}.$$

Hence, by including identity transformation I in our list of rotational symmetries, we can write any number of rotations of $\triangle ABC$ by $120°$ using only three transformations. For this reason, we include the identity transformation as a type of symmetry as well.

Exercises

Use Figure 1 to answer the questions below.

1. Draw all lines of symmetry. Locate the center of rotational symmetry.

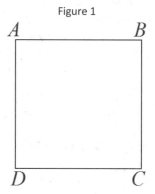
Figure 1

2. Describe all symmetries explicitly.
   a. What kinds are there?

   b. How many are rotations? (Include 360° rotational symmetry, i.e., the identity symmetry.)

   c. How many are reflections?

3. Prove that you have found all possible symmetries.
   a. How many places can vertex A be moved to by some symmetry of the square that you have identified? (Note that the vertex to which you move A by some specific symmetry is known as the image of A under that symmetry. Did you remember the identity symmetry?)

   b. For a given symmetry, if you know the image of A, how many possibilities exist for the image of B?

   c. Verify that there is symmetry for all possible images of A and B.

   d. Using part (b), count the number of possible images of A and B. This is the total number of symmetries of the square. Does your answer match up with the sum of the numbers from Exercise 2 parts (b) and (c)?

Relevant Vocabulary

REGULAR POLYGON: A polygon is *regular* if all sides have equal length and all interior angles have equal measure.

Problem Set

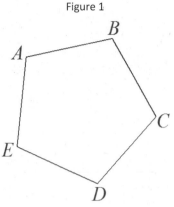

Figure 1

Use Figure 1 to answer Problems 1–3.

1. Draw all lines of symmetry. Locate the center of rotational symmetry.

2. Describe all symmetries explicitly.
   a. What kinds are there?
   b. How many are rotations (including the identity symmetry)?
   c. How many are reflections?

3. Now that you have found the symmetries of the pentagon, consider these questions:
   a. How many places can vertex A be moved to by some symmetry of the pentagon? (Note that the vertex to which you move A by some specific symmetry is known as the image of A under that symmetry. Did you remember the identity symmetry?)
   b. For a given symmetry, if you know the image of A, how many possibilities exist for the image of B?
   c. Verify that there is symmetry for all possible images of A and B.
   d. Using part (b), count the number of possible images of A and B. This is the total number of symmetries of the figure. Does your answer match up with the sum of the numbers from Problem 2 parts (b) and (c)?

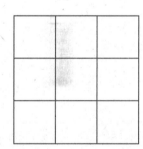

Figure 2

Use Figure 2 to answer Problem 4.

4. Shade exactly two of the nine smaller squares so that the resulting figure has
   a. Only one vertical and one horizontal line of symmetry.
   b. Only two lines of symmetry about the diagonals.
   c. Only one horizontal line of symmetry.
   d. Only one line of symmetry about a diagonal.
   e. No line of symmetry.

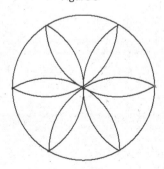

Figure 3

Use Figure 3 to answer Problem 5.

5. Describe all the symmetries explicitly.
   a. How many are rotations (including the identity symmetry)?
   b. How many are reflections?
   c. How could you shade the figure so that the resulting figure only has three possible rotational symmetries (including the identity symmetry)?

Lesson 15: Rotations, Reflections, and Symmetry

6. Decide whether each of the statements is true or false. Provide a counterexample if the answer is false.

   a. If a figure has exactly two lines of symmetry, it has exactly two rotational symmetries (including the identity symmetry).

   b. If a figure has at least three lines of symmetry, it has at least three rotational symmetries (including the identity symmetry).

   c. If a figure has exactly two rotational symmetries (including the identity symmetry), it has exactly two lines of symmetry.

   d. If a figure has at least three rotational symmetries (including the identity symmetry), it has at least three lines of symmetry.

Lesson 16: Translations

Classwork

Exploratory Challenge

In Lesson 4, you completed a construction exercise that resulted in a pair of parallel lines (Problem 1 from the Problem Set). Now we examine an alternate construction.

Construct the line parallel to a given line AB through a given point P.

1. Draw circle P: Center P, radius AB.
2. Draw circle B: Center B, radius AP.
3. Label the intersection of circle P and circle B as Q.
4. Draw $\overleftrightarrow{PQ}$.

Note: Circles P and B intersect in two locations. Pick the intersection Q so that points A and Q are in opposite half-planes of line PB.

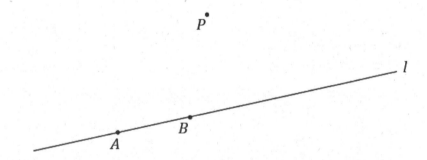

A STORY OF FUNCTIONS

Lesson 16 M1

GEOMETRY

Discussion

To perform a translation, we need to use the previous construction. Let us investigate the definition of translation.

For vector $\vec{AB}$, the *translation along* $\vec{AB}$ is the transformation $T_{\vec{AB}}$ of the plane defined as follows:

1. For any point P on the line AB, $T_{\vec{AB}}(P)$ is the point Q on $\overleftrightarrow{AB}$ so that $\vec{PQ}$ has the same length and the same direction as $\vec{AB}$, and

2. For any point P not on $\overleftrightarrow{AB}$, $T_{\vec{AB}}(P)$ is the point Q obtained as follows. Let l be the line passing through P and parallel to $\overleftrightarrow{AB}$. Let m be the line passing through B and parallel to line AP. The point Q is the intersection of l and m.

Note: The parallel line construction on the previous page shows a quick way to find the point Q in part 2 of the definition of translation.

In the figure to the right, quadrilateral $ABCD$ has been translated the length and direction of vector $\vec{CC'}$. Notice that the distance and direction from each vertex to its corresponding vertex on the image are identical to that of $\vec{CC'}$.

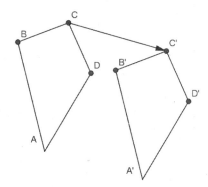

Example 1

Draw the vector that defines each translation below.

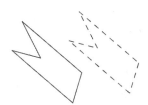 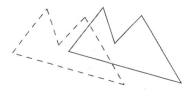 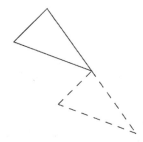

Finding the vector is relatively straightforward. Applying a vector to translate a figure is more challenging. To translate a figure, we must construct parallel lines to the vector through the vertices of the original figure and then find the points on those parallel lines that are the same direction and distance away as given by the vector.

Lesson 16: Translations

A STORY OF FUNCTIONS   Lesson 16   M1

GEOMETRY

Example 2

Use your compass and straightedge to apply $T_{\overrightarrow{AB}}$ to segment P_1P_2.

Note: Use the steps from the Exploratory Challenge *twice* for this question, creating two lines parallel to $\overrightarrow{AB}$: one through P_1 and one through P_2.

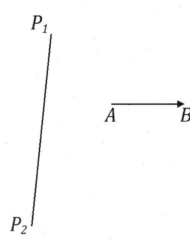

Example 3

Use your compass and straightedge to apply $T_{\overrightarrow{AB}}$ to $\triangle P_1P_2P_3$.

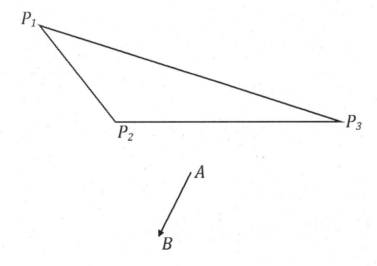

Relevant Vocabulary

PARALLEL: Two lines are *parallel* if they lie in the same plane and do not intersect. Two segments or rays are parallel if the lines containing them are parallel lines.

Lesson 16:   Translations   S.103

A STORY OF FUNCTIONS — Lesson 16 M1

GEOMETRY

Lesson Summary

A translation carries segments onto segments of equal length.

A translation carries angles onto angles of equal measure.

Problem Set

Translate each figure according to the instructions provided.

1. 2 units down and 3 units left

   Draw the vector that defines the translation.

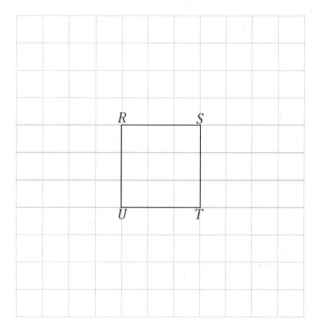

2. 1 unit up and 2 units right

   Draw the vector that defines the translation.

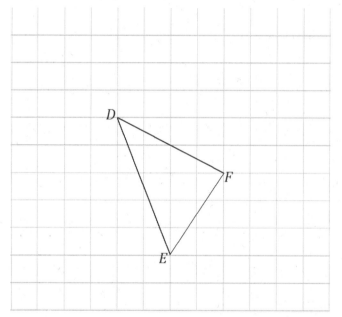

Lesson 16: Translations

3. Use your compass and straightedge to apply $T_{\overrightarrow{AB}}$ to the circle below (center P_1, radius $\overline{P_1P_2}$).

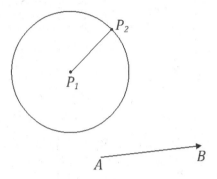

4. Use your compass and straightedge to apply $T_{\overrightarrow{AB}}$ to the circle below.

   Hint: You need to first find the center of the circle. You can use what you learned in Lesson 4 to do this.

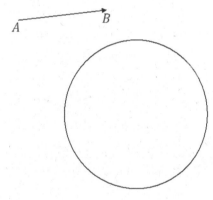

Two classic toothpick puzzles appear below. Solve each puzzle.

5. Each segment on the fish represents a toothpick. Move (translate) exactly three toothpicks and the eye to make the fish swim in the opposite direction. Show the translation vectors needed to move each of the three toothpicks and the eye.

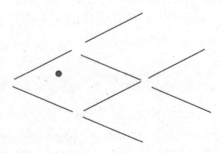

Lesson 16: Translations

6. Again, each segment represents a single toothpick. Move (translate) exactly three toothpicks to make the triangle point downward. Show the translation vectors needed to move each of the three toothpicks.

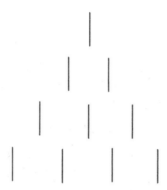

7. Apply $T_{\overrightarrow{GH}}$ to translate $\triangle ABC$.

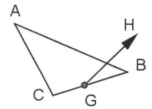

Lesson 17: Characterize Points on a Perpendicular Bisector

Classwork

Opening Exercise

In Lesson 3, you bisected angles, including straight angles. You related the bisection of straight angles in Lesson 3 to the construction of perpendicular bisectors in Lesson 4. Review the process of constructing a perpendicular bisector with the segment below. Then complete the definition of perpendicular lines below your construction.

Use the compass and straightedge construction from Lesson 4.

Two lines are perpendicular if they _____, and if any of the angles formed by the intersection of the lines is a _____ angle. Two segments are perpendicular if the lines containing them are _____.

Discussion

The line you constructed in the Opening Exercise is called the perpendicular bisector of the segment. As you learned in Lesson 14, the perpendicular bisector is also known as the line of reflection of the segment. With a line of reflection, any point on one side of the line (pre-image) is the same distance from the line as its image on the opposite side of the line.

Example 1

Is it possible to find or construct a line of reflection that is *not* a perpendicular bisector of a segment connecting a point on the pre-image to its image? Try to locate a line of reflection between the two figures to the right without constructing any perpendicular bisectors.

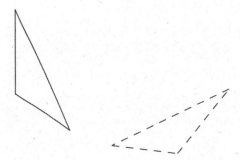

Discussion

Why were your attempts impossible? Look back at the definition of *reflection* from Lesson 14.

> For a line l in the plane, a *reflection across* l is the transformation r_l of the plane defined as follows:
> 1. For any point P on the line l, $r_l(P) = P$, and
> 2. For any point P not on l, $r_l(P)$ is the point Q so that l is the perpendicular bisector of the segment PQ.

The key lies in the use of the term *perpendicular bisector*. For a point P not on l, explain how to construct the point Q so that l is the perpendicular bisector of the segment PQ.

Now, let's think about the problem from another perspective. We have determined that any point on the pre-image figure is the same distance from the line of reflection as its image. Therefore, the two points are equidistant from the point at which the line of reflection (perpendicular bisector) intersects the segment connecting the pre-image point to its image. What about other points on the perpendicular bisector? Are they also equidistant from the pre-image and image points? Let's investigate.

Example 2

Using the same figure from the previous investigation, but with the line of reflection, is it possible to conclude that any point on the perpendicular bisector is equidistant from any pair of pre-image and image points? For example, is $GP = HP$ in the figure? The point P is clearly *not* on the segment connecting the pre-image point G to its image H. How can you be certain that $GP = HP$? If r is the reflection, then $r(G) = H$ and $r(P) = P$. Since r preserves distances, $GP = HP$.

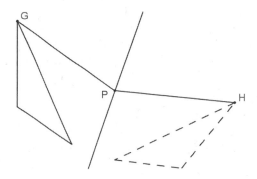

Lesson 17 M1

GEOMETRY

Discussion

We have explored perpendicular bisectors as they relate to reflections and have determined that they are essential to reflections. Are perpendicular lines, specifically, perpendicular bisectors, essential to the other two types of rigid motions: rotations and translations? Translations involve constructing parallel lines (which can certainly be done by constructing perpendiculars but are not essential to constructing parallels). However, perpendicular bisectors play an important role in rotations. In Lesson 13, we found that the intersection of the perpendicular bisectors of two segments connecting pairs of pre-image to image points determined the center of rotation.

Example 3

Find the center of rotation for the transformation below. How are perpendicular bisectors a major part of finding the center of rotation? Why are they essential?

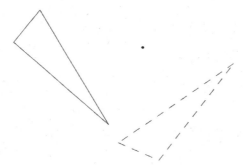

As you explore this figure, also note another feature of rotations. As with all rigid motions, rotations preserve distance. A transformation is said to be distance-preserving (or length-preserving) if the distance between the images of two points is always equal to the distance between the original two points. Which of the statements below is true of the distances in the figure? Justify your response.

1. $AB = A'B'$
2. $AA' = BB'$

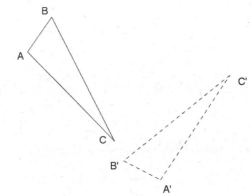

Lesson 17: Characterize Points on a Perpendicular Bisector

Exercises

In each pre-image/image combination below, (a) identify the type of transformation; (b) state whether perpendicular bisectors play a role in constructing the transformation and, if so, what role; and (c) cite an illustration of the distance-preserving characteristic of the transformation (e.g., identify two congruent segments from the pre-image to the image). For the last requirement, you have to label vertices on the pre-image and image.

1.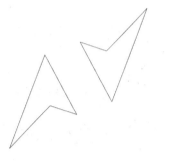

Transformation	Perpendicular Bisectors?	Examples of Distance Preservation

2.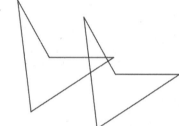

Transformation	Perpendicular Bisectors?	Examples of Distance Preservation

3.

Transformation	Perpendicular Bisectors?	Examples of Distance Preservation

4.

Transformation	Perpendicular Bisectors?	Examples of Distance Preservation

Lesson 17: Characterize Points on a Perpendicular Bisector

5. In the figure to the right, $\overline{GH}$ is a segment of reflection. State and justify two conclusions about distances in this figure. At least one of your statements should refer to perpendicular bisectors.

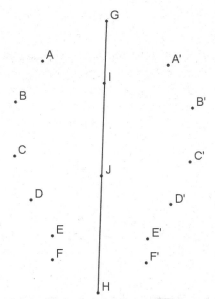

Problem Set

Create/construct two problems involving transformations—one reflection and one rotation—that require the use of perpendicular bisectors. Your reflection problem may require locating the line of reflection or using the line of reflection to construct the image. Your rotation problem should require location of the point of rotation. (Why should your rotation problem *not* require construction of the rotated image?) Create the problems on one page, and construct the solutions on another. Another student will be solving your problems in the next class period.

Lesson 18: Looking More Carefully at Parallel Lines

Classwork

Opening Exercise

Exchange Problem Sets from Lesson 17 with a classmate. Solve the problems posed by your classmate while he or she solves yours. Compare your solutions, and then discuss and resolve any discrepancies. Why were you asked only to locate the point of rotation rather than to rotate a pre-image to obtain the image? How did you use perpendicular bisectors in constructing your solutions?

Discussion

We say that two lines are *parallel* if they lie in the same plane and do not intersect. Two segments or rays are parallel if the lines containing them are parallel.

Example 1

Why is the phrase *in the plane* critical to the definition of parallel lines? Explain and illustrate your reasoning.

In Lesson 7, we recalled some basic facts learned in earlier grades about pairs of lines and angles created by a transversal to those lines. One of those basic facts is the following:

> Suppose a transversal intersects a pair of lines. The lines are parallel if and only if a pair of alternate interior angles are equal in measure.

Our goal in this lesson is to prove this theorem using basic rigid motions, geometry assumptions, and a geometry assumption we introduce in this lesson called the *parallel postulate*. Of all of the geometry assumptions we have given so far, the parallel postulate gets a special name because of the special role it played in the history of mathematics. (Euclid included a version of the parallel postulate in his books, and for 2,000 years people tried to show that it was not a necessary assumption. Not only did it turn out that the assumption was necessary for Euclidean geometry, but study of the parallel postulate led to the creation of non-Euclidean geometries.)

The basic fact above really has two parts, which we prove separately:

1. Suppose a transversal intersects a pair of lines. If two alternate interior angles are equal in measure, then the pair of lines are parallel.
2. Suppose a transversal intersects a pair of lines. If the lines are parallel, then the pair of alternate interior angles are equal in measure.

The second part turns out to be an equivalent form of the parallel postulate. To build up to the theorem, first we need to do a construction.

Example 2

Given a line l and a point P not on the line, follow the steps below to rotate l by 180° to a line l' that passes through P:

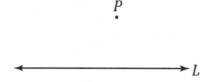

a. Label any point A on l.

b. Find the midpoint of segment AP using a ruler. (Measure the length of segment AP, and locate the point that is distance $\frac{AP}{2}$ from A between A and P.) Label the midpoint C.

c. Perform a 180° rotation around center C. To quickly find the image of l under this rotation by hand:
   i. Pick another point B on l.
   ii. Draw $\overleftrightarrow{CB}$.
   iii. Draw circle: center C, radius CB.
   iv. Label the other point where the circle intersects $\overleftrightarrow{CB}$ by Q.
   v. Draw $\overleftrightarrow{PQ}$.

d. Label the image of the rotation by 180° of l by $l' = R_{C,180}(l)$.

How does your construction relate to the geometry assumption stated above to rotations? Complete the statement below to clarify your observations:

$R_{C,180}$ is a 180° _____ around C. Rotations preserve _____ ; therefore $R_{C,180}$, maps the line l to the line _____. What is $R_{C,180}(A)$? _____

Example 3

The lines l and l' in the construction certainly look parallel, but we do not have to rely on looks.

Claim: In the construction, l is parallel to l'.

PROOF: We show that assuming they are not parallel leads to a contradiction. If they are not parallel, then they must intersect somewhere. Call that point X. Since X is on l', it must be the image of some point S on l under the $R_{C,180}$ rotation, (i.e., $R_{C,180}(S) = X$). Since $R_{C,180}$ is a $180°$ rotation, S and X must be the endpoints of a diameter of a circle that has center C. In particular, $\overleftrightarrow{SX}$ must contain C. Since S is a point on l, and X is a different point on l (it was the intersection of both lines), we have that $l = \overleftrightarrow{SX}$ because there is only one line through two points. But $\overleftrightarrow{SX}$ also contains C, which means that l contains C. However, C was constructed so that it was not on l. This is absurd.

There are only two possibilities for any two distinct lines l and l' in a plane: either the lines are parallel, or they are not parallel. Since assuming the lines were not parallel led to a false conclusion, the only possibility left is that l and l' were parallel to begin with.

Example 4

The construction and claim together implies the following theorem.

THEOREM: Given a line l and a point P not on the line, then there exists line l' that contains P and is parallel to l.

This is a theorem we have justified before using compass and straightedge constructions, but now we see it follows directly from basic rigid motions and our geometry assumptions.

Example 5

We are now ready to prove the first part of the basic fact above. We have two lines, l and l', and all we know is that a transversal $\overleftrightarrow{AP}$ intersects l and l' such that a pair of alternate interior angles are equal in measure. (In the picture below, we are assuming $m\angle QPA = m\angle BAP$.)

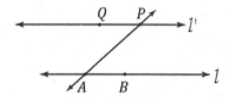

Let C be the midpoint of $\overline{AP}$. What happens if you rotate 180° around the center C? Is there enough information to show that $R_{C,180}(l) = l'$?

a. What is the image of the segment AP?

b. In particular, what is the image of the point A?

c. Why are the points Q and $R_{C,180}(B)$ on the same side of $\overleftrightarrow{AP}$?

d. What is the image of $R_{C,180}(\angle BAP)$? Is it $\angle QPA$? Explain why.

e. Why is $R_{C,180}(l) = l'$?

We have just proven that a rotation by 180° takes l to l'. By the claim in Example 3, lines l and l' must be parallel, which is summarized below.

THEOREM: Suppose a transversal intersects a pair of lines. If a pair of alternate interior angles are equal in measure, then the pair of lines are parallel.

Discussion

In Example 5, suppose we had used a different rotation to construct a line parallel to l that contains P. Such constructions are certainly plentiful. For example, for every other point D on l, we can find the midpoint of segment PD and use the construction in Example 2 to construct a different 180° rotation around a different center such that the image of the line l is a parallel line through the point P. Are any of these parallel lines through P different? In other words,

Can we draw a line other than the line l' through P that never meets l?

The answer may surprise you; it stumped mathematicians and physicists for centuries. In nature, the answer is that it is sometimes possible and sometimes not. This is because there are places in the universe (near massive stars, for example) where the model geometry of space is not plane-like or flat but is actually quite curved. To rule out these other types of strange but beautiful geometries, we must assume that the answer to the previous question is only one line. That choice becomes one of our geometry assumptions:

(Parallel Postulate) *Through a given external point there is at most one line parallel to a given line.*

In other words, we assume that for any point P in the plane not lying on a line ℓ, every line in the plane that contains P intersects ℓ except at most one line—the one we call *parallel* to ℓ.

Example 6

We can use the parallel postulate to prove the second part of the basic fact.

THEOREM: Suppose a transversal intersects a pair of lines. If the pair of lines are parallel, then the pair of alternate interior angles are equal in measure.

PROOF: Suppose that a transversal $\overleftrightarrow{AP}$ intersects line l at A and l' at P, pick and label another point B on l, and choose a point Q on l' on the opposite side of $\overleftrightarrow{AP}$ as B. The picture might look like the figure below:

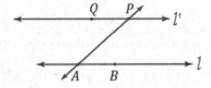

Let C be the midpoint of $\overline{AP}$, and apply a rotation by 180° around the center C. As in previous discussions, the image of l is the line $R_{C,180}(l)$, which is parallel to l and contains point P. Since l' and $R_{C,180}(l)$ are both parallel to l and contain P, by the parallel postulate, they must be the same line: $R_{C,180}(l) = l'$. In particular, $R_{C,180}(\angle BAP) = \angle QPA$. Since rotations preserve angle measures, $m\angle BAP = m\angle QPA$, which was what we needed to show.

Discussion

It is important to point out that, although we only proved the alternate interior angles theorem, the same sort of proofs can be done in the exact same way to prove the corresponding angles theorem and the interior angles theorem. Thus, all of the proofs we have done so far (in class and in the Problem Sets) that use these facts are really based, in part, on our assumptions about rigid motions.

Example 7

We end this lesson with a theorem that we just state but can be easily proved using the parallel postulate.

THEOREM: If three distinct lines l_1, l_2, and l_3 in the plane have the property that $l_1 \parallel l_2$ and $l_2 \parallel l_3$, then $l_1 \parallel l_3$. (In proofs, this can be written as, "If two lines are parallel to the same line, then they are parallel to each other.")

Relevant Vocabulary

PARALLEL: Two lines are *parallel* if they lie in the same plane and do not intersect. Two segments or rays are parallel if the lines containing them are parallel lines.

TRANSVERSAL: Given a pair of lines l and m in a plane, a third line t is a *transversal* if it intersects l at a single point and intersects m at a single but different point.

The definition of transversal rules out the possibility that any two of the lines l, m, and t are the same line.

ALTERNATE INTERIOR ANGLES: Let line t be a transversal to lines l and m such that t intersects l at point P and intersects m at point Q. Let R be a point on l and S be a point on m such that the points R and S lie in opposite half planes of t. Then the $\angle RPQ$ and the $\angle PQS$ are called *alternate interior angles* of the transversal t with respect to m and l.

CORRESPONDING ANGLES: Let line t be a transversal to lines l and m. If $\angle x$ and $\angle y$ are alternate interior angles, and $\angle y$ and $\angle z$ are vertical angles, then $\angle x$ and $\angle z$ are *corresponding angles*.

Problem Set

Notice that we are frequently asked two types of questions about parallel lines. If we are told that two lines are parallel, then we may be required to use this information to prove the congruence of two angles (corresponding, alternate interior, etc.). On the other hand, if we are given the fact that two angles are congruent (or perhaps supplementary), we may have to prove that two lines are parallel.

1. In the figure, $\overline{AL} \parallel \overline{BM}$, $\overline{AL} \perp \overline{CF}$, and $\overline{GK} \perp \overline{BM}$. Prove that $\overline{CF} \parallel \overline{GK}$.

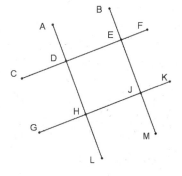

2. Given that $\angle B$ and $\angle C$ are supplementary and $m\angle A = m\angle C$, prove that $\overline{AD} \parallel \overline{BC}$.

3. Mathematicians state that *if a transversal to two parallel lines is perpendicular to one of the lines, then it is perpendicular to the other.* Prove this statement. (Include a labeled drawing with your proof.)

4. In the figure, $\overline{AB} \parallel \overline{CD}$ and $\overline{EF} \parallel \overline{GH}$. Prove that $m\angle AFE = m\angle DKH$.

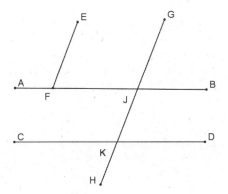

Lesson 18: Looking More Carefully at Parallel Lines

5. In the figure, ∠E and ∠AFE are complementary, and ∠C and ∠BDC are complementary. Prove that $\overline{AE} \parallel \overline{CB}$.

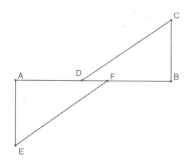

6. Given a line l and a point P not on the line, the following directions can be used to draw a line m perpendicular to the line l through the point P based upon a rotation by 180°:

   a. Pick and label a point A on the line l so that the circle (center P, radius AP) intersects l twice.

   b. Use a protractor to draw a perpendicular line n through the point A (by constructing a 90° angle).

   c. Use the directions in Example 2 to construct a parallel line m through the point P.

   Do the construction. Why is the line m perpendicular to the line l in the figure you drew? Why is the line m the only perpendicular line to l through P?

Problems 7–10 all refer to the figure to the right. The exercises are otherwise unrelated to each other.

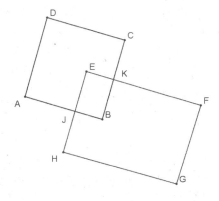

7. $\overline{AD} \parallel \overline{BC}$ and ∠EJB is supplementary to ∠JBK. Prove that $\overline{AD} \parallel \overline{JE}$.

8. $\overline{AD} \parallel \overline{FG}$ and $\overline{EJ} \parallel \overline{FG}$. Prove that ∠DAJ and ∠EJA are supplementary.

9. $m\angle C = m\angle G$ and ∠B is supplementary to ∠G. Prove that $\overline{DC} \parallel \overline{AB}$.

10. $\overline{AB} \parallel \overline{EF}$, $\overline{EF} \perp \overline{CB}$, and ∠EKC is supplementary to ∠KCD. Prove that $\overline{AB} \parallel \overline{DC}$.

Lesson 19: Construct and Apply a Sequence of Rigid Motions

Classwork

Opening

We have been using the idea of congruence already (but in a casual and unsystematic way). In Grade 8, we introduced and experimented with concepts around congruence through *physical models, transparencies, or geometry software*. Specifically, we had to

> *(1) Understand that a two-dimensional figure is congruent to another if the second can be obtained from the first by a sequence of rotations, reflections, and translations; and (2) describe a sequence that exhibits the congruence between two congruent figures.* **(8.G.A.2)**

As with so many other concepts in high school Geometry, congruence is familiar, but we now study it with greater precision and focus on the language with which we discuss it.

Let us recall some facts related to congruence that appeared previously in this unit.

1. We observed that rotations, translations, and reflections—and thus all rigid motions—preserve the lengths of segments and the measures of angles. We think of two segments (respectively, angles) as the *same* in an important respect if they have the same length (respectively, degree measure), and thus, sameness of these objects relating to measure is well characterized by the existence of a rigid motion mapping one thing to another. Defining *congruence* by means of rigid motions extends this notion of sameness to arbitrary figures, while clarifying the meaning in an articulate way.

2. We noted that a symmetry is a rigid motion that carries a figure to itself.

So how do these facts about rigid motions and symmetry relate to congruence? We define two figures in the plane as congruent if there exists a finite composition of basic rigid motions that maps one figure onto the other.

It might seem easy to equate two figures being congruent to having *same size and same shape*. The phrase *same size and same shape* has intuitive meaning and helps to paint a mental picture, but it is not a definition. As in a court of law, to establish guilt it is not enough to point out that the defendant looks like a sneaky, unsavory type. We need to point to exact pieces of evidence concerning the specific charges. It is also not enough that the defendant did something bad. It must be a violation of a specific law. Same size and same shape is on the level of, "He looks like a sneaky, bad guy who deserves to be in jail."

It is also not enough to say that they are alike in all respects except position in the plane. We are saying that there is some particular rigid motion that carries one to another. Almost always, when we use congruence in an explanation or proof, we need to refer to the rigid motion. To show that two figures are congruent, we only need to show that there is a transformation that maps one directly onto the other. However, once we know that there is a transformation, then we know that there are actually many such transformations, and it can be useful to consider more than one. We see this when discussing the symmetries of a figure. A symmetry is nothing other than a congruence of an object with itself.

A figure may have many different rigid motions that map it onto itself. For example, there are six different rigid motions that take one equilateral triangle with side length 1 to another such triangle. Whenever this occurs, it is because of a symmetry in the objects being compared.

Lastly, we discuss the relationship between *congruence* and *correspondence*. A correspondence between two figures is a function from the parts of one figure to the parts of the other, with no requirements concerning same measure or existence of rigid motions. If we have rigid motion T that takes one figure to another, then we have a correspondence between the parts. For example, if the first figure contains segment AB, then the second includes a corresponding segment $T(A)T(B)$. But we do not need to have a congruence to have a correspondence. We might list the parts of one figure and pair them with the parts of another. With two triangles, we might match vertex to vertex. Then the sides and angles in the first have corresponding parts in the second. But being able to set up a correspondence like this does not mean that there is a rigid motion that produces it. The sides of the first might be paired with sides of different length in the second. Correspondence in this sense is important in triangle similarity.

Discussion

We now examine a figure being mapped onto another through a composition of rigid motions.

To map $\triangle PQR$ to $\triangle XYZ$ here, we first rotate $\triangle PQR$ 120° ($R_{D,120°}$) around the point, D. Then reflect the image ($r_{\overline{EF}}$) across $\overline{EF}$. Finally, translate the second image ($T_{\vec{v}}$) along the given vector to obtain $\triangle XYZ$. Since each transformation is a rigid motion, $\triangle PQR \cong \triangle XYZ$. We use function notation to describe the composition of the rotation, reflection, and translation:

$$T_{\vec{v}}\left(r_{\overline{EF}}\left(R_{D,120°}(\triangle PQR)\right)\right) = \triangle XYZ.$$

Notice that (as with all composite functions) the innermost function/transformation (the rotation) is performed first, and the outermost (the translation) last.

Example 1

i. Draw and label a $\triangle PQR$ in the space below.
ii. Use your construction tools to apply one of each of the rigid motions we have studied to it in a sequence of your choice.
iii. Use function notation to describe your chosen composition here. Label the resulting image as $\triangle XYZ$:
   _____
iv. Complete the following sentences: (Some blanks are single words; others are phrases.)

$\triangle PQR$ is _____ to $\triangle XYZ$ because _____ map point P to point X, point Q to point Y, and point R to point Z. Rigid motions map segments onto _____ and angles onto _____.

Example 2

On a separate piece of paper, trace the series of figures in your composition but do *NOT* include the center of rotation, the line of reflection, or the vector of the applied translation.

Swap papers with a partner, and determine the composition of transformations your partner used. Use function notation to show the composition of transformations that renders △ $PQR \cong$ △ XYZ.

Problem Set

1. Use your understanding of congruence to explain why a triangle cannot be congruent to a quadrilateral.
   a. Why can't a triangle be congruent to a quadrilateral?
   b. Why can't an isosceles triangle be congruent to a triangle that is not isosceles?

2. Use the figures below to answer each question:
   a. $\triangle ABD \cong \triangle CDB$. What rigid motion(s) maps $\overline{CD}$ onto $\overline{AB}$? Find two possible solutions.

   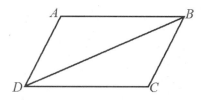

   b. All of the smaller triangles are congruent to each other. What rigid motion(s) map $\overline{ZB}$ onto $\overline{AZ}$? Find two possible solutions.

   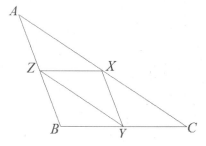

Lesson 20: Applications of Congruence in Terms of Rigid Motions

Classwork

Opening

Every congruence gives rise to a correspondence.

Under our definition of congruence, when we say that one figure is congruent to another, we mean that there is a rigid motion that maps the first onto the second. That rigid motion is called a congruence.

Recall the Grade 7 definition: *A correspondence between two triangles is a pairing of each vertex of one triangle with one and only one vertex of the other triangle.* When reasoning about figures, it is useful to be able to refer to corresponding parts (e.g., sides and angles) of the two figures. We look at one part of the first figure and compare it to the corresponding part of the other. Where does a correspondence come from? We might be told by someone how to make the vertices correspond. Conversely, we might make our own correspondence by matching the parts of one triangle with the parts of another triangle based on appearance. Finally, if we have a congruence between two figures, the congruence gives rise to a correspondence.

A rigid motion F always produces a one-to-one correspondence between the points in a figure (the *pre-image*) and points in its image. If P is a point in the figure, then the corresponding point in the image is $F(P)$. A rigid motion also maps each part of the figure to a corresponding part of the image. As a result, *corresponding parts of congruent figures are congruent* since the very same rigid motion that makes a congruence between the figures also makes a congruence between each part of the figure and the corresponding part of the image.

In proofs, we frequently refer to the fact that corresponding angles, sides, or parts of congruent triangles are congruent. This is simply a repetition of the definition of congruence. If △ ABC is congruent to △ DEG because there is a rigid motion F such that $F(A) = D$, $F(B) = E$, and $F(C) = G$, then $\overline{AB}$ is congruent to $\overline{DE}$, △ ABC is congruent to △ DEG, and so forth because the rigid motion F takes $\overline{AB}$ to $\overline{DE}$ and $\angle BAC$ to $\angle EDF$.

There are correspondences that do not come from congruences.

The sides (and angles) of two figures might be compared even when the figures are not congruent. For example, a carpenter might want to know if two windows in an old house are the same, so the screen for one could be interchanged with the screen for the other. He might list the parts of the first window and the analogous parts of the second, thus making a correspondence between the parts of the two windows. Checking part by part, he might find that the angles in the frame of one window are slightly different from the angles in the frame of the other, possibly because the house has tilted slightly as it aged. He has used a correspondence to help describe the differences between the windows not to describe a congruence.

In general, given any two triangles, one could make a table with two columns and three rows and then list the vertices of the first triangle in the first column and the vertices of the second triangle in the second column in a random way. This would create a correspondence between the triangles, though generally not a very useful one. No one would expect a random correspondence to be very useful, but it is a correspondence nonetheless.

Later, when we study similarity, we find that it is very useful to be able to set up correspondences between triangles despite the fact that the triangles are not congruent. Correspondences help us to keep track of which part of one figure we are comparing to that of another. It makes the rules for associating part to part explicit and systematic so that other people can plainly see what parts go together.

Lesson 20: Applications of Congruence in Terms of Rigid Motions

A STORY OF FUNCTIONS
Lesson 20 M1
GEOMETRY

Discussion

Let's review function notation for rigid motions.

a. To name a translation, we use the symbol $T_{\overrightarrow{AB}}$. We use the letter T to signify that we are referring to a translation and the letters A and B to indicate the translation that moves each point in the direction from A to B along a line parallel to line AB by distance AB. The image of a point P is denoted $T_{\overrightarrow{AB}}(P)$. Specifically, $T_{\overrightarrow{AB}}(A) = B$.

b. To name a reflection, we use the symbol r_l, where l is the line of reflection. The image of a point P is denoted $r_l(P)$. In particular, if A is a point on l, $r_l(A) = A$. For any point P, line l is the perpendicular bisector of segment $Pr_l(P)$.

c. To name a rotation, we use the symbol $R_{C,x°}$ to remind us of the word *rotation*. C is the center point of the rotation, and x represents the degree of the rotation counterclockwise around the center point. Note that a positive degree measure refers to a counterclockwise rotation, while a negative degree measure refers to a clockwise rotation.

Example 1

In each figure below, the triangle on the left has been mapped to the one on the right by a 240° rotation about P. Identify all six pairs of corresponding parts (vertices and sides).

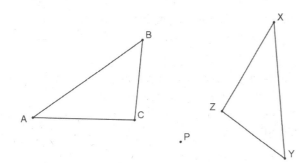

Corresponding Vertices	Corresponding Sides

What rigid motion mapped △ABC onto △XYZ? Write the transformation in function notation.

Example 2

Given a triangle with vertices A, B, and C, list all the possible correspondences of the triangle with itself.

Example 3

Give an example of two quadrilaterals and a correspondence between their vertices such that (a) corresponding sides are congruent, but (b) corresponding angles are not congruent.

A STORY OF FUNCTIONS Lesson 20 M1
GEOMETRY

Problem Set

1. Given two triangles, one with vertices A, B, and C, and the other with vertices X, Y, and Z, there are six different correspondences of the first with the second.

   a. One such correspondence is the following:
   $$A \to Z$$
   $$B \to X$$
   $$C \to Y$$
   Write the other five correspondences.

   b. If all six of these correspondences come from congruences, then what can you say about $\triangle ABC$?

   c. If two of the correspondences come from congruences, but the others do not, then what can you say about $\triangle ABC$?

   d. Why can there be no two triangles where three of the correspondences come from congruences, but the others do not?

2. Give an example of two triangles and a correspondence between them such that (a) all three corresponding angles are congruent, but (b) corresponding sides are not congruent.

3. Give an example of two triangles and a correspondence between their vertices such that (a) one angle in the first is congruent to the corresponding angle in the second, and (b) two sides of the first are congruent to the corresponding sides of the second, but (c) the triangles themselves are not congruent.

4. Give an example of two quadrilaterals and a correspondence between their vertices such that (a) all four corresponding angles are congruent, and (b) two sides of the first are congruent to two sides of the second, but (c) the two quadrilaterals are not congruent.

5. A particular rigid motion, M, takes point P as input and gives point P' as output. That is, $M(P) = P'$. The same rigid motion maps point Q to point Q'. Since rigid motions preserve distance, is it reasonable to state that $PP' = QQ'$? Does it matter which type of rigid motion M is? Justify your response for each of the three types of rigid motion. Be specific. If it is indeed the case, for some class of transformations, that $PP' = QQ'$ is true for all P and Q, explain why. If not, offer a counterexample.

Lesson 20: Applications of Congruence in Terms of Rigid Motions

Lesson 21: Correspondence and Transformations

Classwork

Opening Exercise

The figure to the right represents a rotation of △ ABC 80° around vertex C. Name the triangle formed by the image of △ ABC. Write the rotation in function notation, and name all corresponding angles and sides.

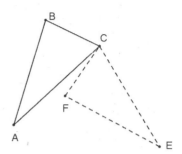

Discussion

In the Opening Exercise, we explicitly showed a single rigid motion, which mapped every side and every angle of △ ABC onto △ EFC. Each corresponding pair of sides and each corresponding pair of angles was congruent. When each side and each angle on the pre-image maps onto its corresponding side or angle on the image, the two triangles are congruent. Conversely, if two triangles are congruent, then each side and angle on the pre-image is congruent to its corresponding side or angle on the image.

A STORY OF FUNCTIONS Lesson 21 M1
GEOMETRY

Example

$ABCD$ is a square, and $\overline{AC}$ is one diagonal of the square. $\triangle ABC$ is a reflection of $\triangle ADC$ across segment AC. Complete the table below, identifying the missing corresponding angles and sides.

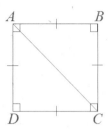

Corresponding Angles	Corresponding Sides
$\angle BAC \rightarrow$	$\overline{AB} \rightarrow$
$\angle ABC \rightarrow$	$\overline{BC} \rightarrow$
$\angle BCA \rightarrow$	$\overline{AC} \rightarrow$

a. Are the corresponding sides and angles congruent? Justify your response.

b. Is $\triangle ABC \cong \triangle ADC$? Justify your response.

Lesson 21: Correspondence and Transformations

A STORY OF FUNCTIONS

Lesson 21   M1

GEOMETRY

Exercises

Each exercise below shows a sequence of rigid motions that map a pre-image onto a final image. Identify each rigid motion in the sequence, writing the composition using function notation. Trace the congruence of each set of corresponding sides and angles through all steps in the sequence, proving that the pre-image is congruent to the final image by showing that every side and every angle in the pre-image maps onto its corresponding side and angle in the image. Finally, make a statement about the congruence of the pre-image and final image.

1.

Sequence of Rigid Motions (2)	
Composition in Function Notation	
Sequence of Corresponding Sides	
Sequence of Corresponding Angles	
Triangle Congruence Statement	

2.

Sequence of Rigid Motions (2)	
Composition in Function Notation	
Sequence of Corresponding Sides	
Sequence of Corresponding Angles	
Triangle Congruence Statement	

Lesson 21:   Correspondence and Transformations

S.131

3.

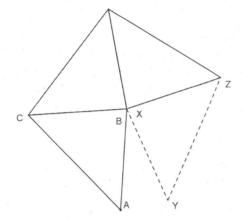

Sequence of Rigid Motions (2)	
Composition in Function Notation	
Sequence of Corresponding Sides	
Sequence of Corresponding Angles	
Triangle Congruence Statement	

Problem Set

1. Exercise 3 mapped △ ABC onto △ YXZ in three steps. Construct a fourth step that would map △ YXZ back onto △ ABC.

2. Explain triangle congruence in terms of rigid motions. Use the terms *corresponding sides* and *corresponding angles* in your explanation.

This page intentionally left blank

Lesson 22: Congruence Criteria for Triangles—SAS

Classwork

Opening Exercise

Answer the following question. Then discuss your answer with a partner.

Do you think it is possible to know whether there is a rigid motion that takes one triangle to another without actually showing the particular rigid motion? Why or why not?

_____

_____

_____

Discussion

It is true that we do not need to show the rigid motion to be able to know that there is one. We are going to show that there are criteria that refer to a few parts of the two triangles and a correspondence between them that guarantee congruency (i.e., existence of rigid motion). We start with the Side-Angle-Side (SAS) criteria.

SIDE-ANGLE-SIDE TRIANGLE CONGRUENCE CRITERIA (SAS): Given two triangles $\triangle ABC$ and $\triangle A'B'C'$ so that $AB = A'B'$ (Side), $m\angle A = m\angle A'$ (Angle), and $AC = A'C'$ (Side). Then the triangles are congruent.

The steps below show the most general case for determining a congruence between two triangles that satisfy the SAS criteria. Note that not all steps are needed for every pair of triangles. For example, sometimes the triangles already share a vertex. Sometimes a reflection is needed, sometimes not. It is important to understand that we can always use some or all of the steps below to determine a congruence between the two triangles that satisfies the SAS criteria.

PROOF: Provided the two distinct triangles below, assume $AB = A'B'$ (Side), $m\angle A = m\angle A'$ (Angle), and $AC = A'C'$ (Side).

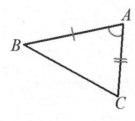 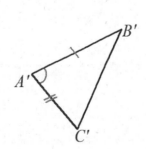

A STORY OF FUNCTIONS

Lesson 22 M1

GEOMETRY

By our definition of congruence, we have to find a composition of rigid motions that maps $\triangle A'B'C'$ to $\triangle ABC$. We must find a congruence F so that $F(\triangle A'B'C') = \triangle ABC$. First, use a translation T to map a common vertex.

Which two points determine the appropriate vector?

Can any other pair of points be used? _____ Why or why not?
_____

State the vector in the picture below that can be used to translate $\triangle A'B'C'$. _____

Using a dotted line, draw an intermediate position of $\triangle A'B'C'$ as it moves along the vector:

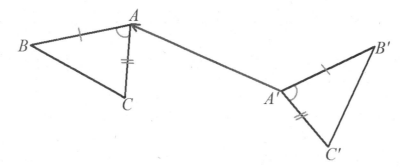

After the translation (below), $T_{\overrightarrow{A'A}}(\triangle A'B'C')$ shares one vertex with $\triangle ABC$, A. In fact, we can say

$T_{\overrightarrow{A'A}}(\triangle \underline{\qquad}) = \triangle \underline{\qquad}$.

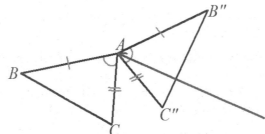

Next, use a clockwise rotation $R_{\angle CAC''}$ to bring the side $\overline{AC''}$ to $\overline{AC}$ (or a counterclockwise rotation to bring $\overline{AB''}$ to $\overline{AB}$).

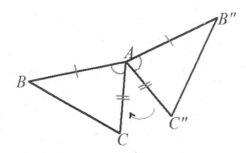

S.136 Lesson 22: Congruence Criteria for Triangles—SAS

A rotation of appropriate measure maps $\overrightarrow{AC''}$ to $\overrightarrow{AC}$, but how can we be sure that vertex C'' maps to C? Recall that part of our assumption is that the lengths of sides in question are equal, ensuring that the rotation maps C'' to C. ($AC = AC''$; the translation performed is a rigid motion, and thereby did not alter the length when $\overline{AC'}$ became $\overline{AC''}$.)

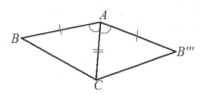

After the rotation $R_{\angle CAC''}(\triangle AB''C'')$, a total of two vertices are shared with $\triangle ABC$, A and C. Therefore,

Finally, if B''' and B are on opposite sides of the line that joins AC, a reflection $r_{\overline{AC}}$ brings B''' to the same side as B.

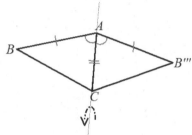

Since a reflection is a rigid motion and it preserves angle measures, we know that $m\angle B'''AC = m\angle BAC$ and so $\overrightarrow{AB'''}$ maps to $\overrightarrow{AB}$. If, however, $\overrightarrow{AB'''}$ coincides with $\overrightarrow{AB}$, can we be certain that B''' actually maps to B? We can, because not only are we certain that the rays coincide but also by our assumption that $AB = AB'''$. (Our assumption began as $AB = A'B'$, but the translation and rotation have preserved this length now as AB'''.) Taken together, these two pieces of information ensure that the reflection over $\overline{AC}$ brings B''' to B.

Another way to visually confirm this is to draw the marks of the _____ construction for $\overline{AC}$.

Write the transformations used to correctly notate the congruence (the composition of transformations) that take $\triangle A'B'C' \cong \triangle ABC$:

F   _____

G   _____

H   _____

We have now shown a sequence of rigid motions that takes $\triangle A'B'C'$ to $\triangle ABC$ with the use of just three criteria from each triangle: two sides and an included angle. Given any two distinct triangles, we could perform a similar proof. There is another situation when the triangles are not distinct, where a modified proof is needed to show that the triangles map onto each other. Examine these below. Note that when using the Side-Angle-Side triangle congruence criteria as a reason in a proof, you need only state the congruence and SAS.

Lesson 22: Congruence Criteria for Triangles—SAS

A STORY OF FUNCTIONS Lesson 22 M1
GEOMETRY

Example

What if we had the SAS criteria for two triangles that were not distinct? Consider the following two cases. How would the transformations needed to demonstrate congruence change?

Case	Diagram	Transformations Needed
Shared Side		
Shared Vertex		

Exercises

1. Given: Triangles with a pair of corresponding sides of equal length and a pair of included angles of equal measure. Sketch and label three phases of the sequence of rigid motions that prove the two triangles to be congruent.

S.138 Lesson 22: Congruence Criteria for Triangles—SAS

Justify whether the triangles meet the SAS congruence criteria; explicitly state which pairs of sides or angles are congruent and why. If the triangles do meet the SAS congruence criteria, describe the rigid motion(s) that would map one triangle onto the other.

2. Given: $m\angle LNM = m\angle LNO$, $MN = ON$

   Do $\triangle LNM$ and $\triangle LNO$ meet the SAS criteria?

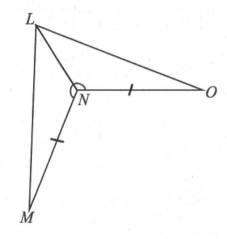

3. Given: $m\angle HGI = m\angle JIG$, $HG = JI$

   Do $\triangle HGI$ and $\triangle JIG$ meet the SAS criteria?

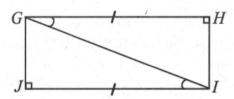

4. Is it true that we could also have proved $\triangle HGI$ and $\triangle JIG$ meet the SAS criteria if we had been given that $\angle HGI \cong \angle JIG$ and $\overline{HG} \cong \overline{JI}$? Explain why or why not.

Lesson 22: Congruence Criteria for Triangles—SAS

A STORY OF FUNCTIONS   Lesson 22 M1
                                                                                           GEOMETRY

Problem Set

Justify whether the triangles meet the SAS congruence criteria; explicitly state which pairs of sides or angles are congruent and why. If the triangles do meet the SAS congruence criteria, describe the rigid motion(s) that would map one triangle onto the other.

1. Given: $\overline{AB} \parallel \overline{CD}$, and $AB = CD$
   Do $\triangle ABD$ and $\triangle CDB$ meet the SAS criteria?

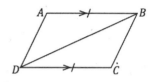

2. Given: $m\angle R = 25°$, $RT = 7"$, $SU = 5"$, and $ST = 5"$
   Do $\triangle RSU$ and $\triangle RST$ meet the SAS criteria?

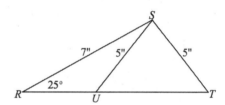

3. Given: $\overline{KM}$ and $\overline{JN}$ bisect each other
   Do $\triangle JKL$ and $\triangle NML$ meet the SAS criteria?

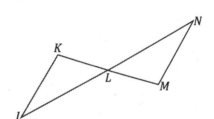

4. Given: $m\angle 1 = m\angle 2$, and $BC = DC$
   Do $\triangle ABC$ and $\triangle ADC$ meet the SAS criteria?

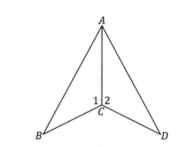

5. Given: $\overline{AE}$ bisects angle $\angle BCD$, and $BC = DC$
   Do $\triangle CAB$ and $\triangle CAD$ meet the SAS criteria?

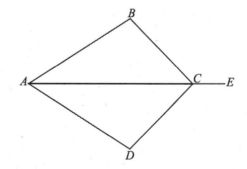

S.140   Lesson 22:   Congruence Criteria for Triangles—SAS

6. Given: $\overline{SU}$ and $\overline{RT}$ bisect each other

   Do △ SVR and △ UVT meet the SAS criteria?

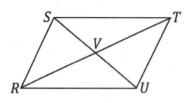

7. Given: $JM = KL$, $\overline{JM} \perp \overline{ML}$, and $\overline{KL} \perp \overline{ML}$

   Do △ JML and △ KLM meet the SAS criteria?

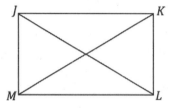

8. Given: $\overline{BF} \perp \overline{AC}$, and $\overline{CE} \perp \overline{AB}$

   Do △ BED and △ CFD meet the SAS criteria?

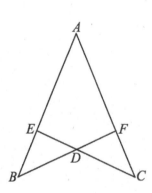

9. Given: $m\angle VXY = m\angle VYX$

   Do △ VXW and △ VYZ meet the SAS criteria?

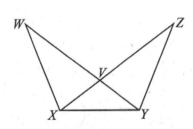

10. Given: △ RST is isosceles, with $RS = RT$, and $SY = TZ$

    Do △ RSY and △ RTZ meet the SAS criteria?

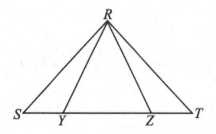

This page intentionally left blank

Lesson 23: Base Angles of Isosceles Triangles

Classwork

Opening Exercise

Describe the additional piece of information needed for each pair of triangles to satisfy the SAS triangle congruence criteria.

a. Given: $AB = DC$

   _____

   Prove: $\triangle ABC \cong \triangle DCB$

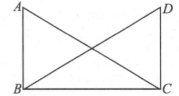

b. Given: $AB = RS$
   $\overline{AB} \parallel \overline{RS}$

   _____

   Prove: $\triangle ABC \cong \triangle RST$

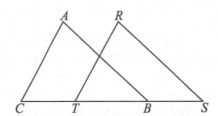

Exploratory Challenge

Today we examine a geometry fact that we already accept to be true. We are going to prove this known fact in two ways: (1) by using transformations and (2) by using SAS triangle congruence criteria.

Here is isosceles triangle ABC. We accept that an isosceles triangle, which has (at least) two congruent sides, also has congruent base angles.

Label the congruent angles in the figure.

Now we prove that the base angles of an isosceles triangle are always congruent.

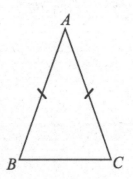

Prove Base Angles of an Isosceles are Congruent: Transformations

Given: Isosceles △ ABC, with AB = AC

Prove: m∠B = m∠C

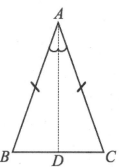

Construction: Draw the angle bisector $\overrightarrow{AD}$ of ∠A, where D is the intersection of the bisector and $\overline{BC}$. We need to show that rigid motions maps point B to point C and point C to point B.

Let r be the reflection through $\overleftrightarrow{AD}$. Through the reflection, we want to demonstrate two pieces of information that map B to point C and vice versa: (1) $\overrightarrow{AB}$ maps to $\overrightarrow{AC}$, and (2) AB = AC.

Since A is on the line of reflection, $\overleftrightarrow{AD}$, $r(A) = A$. Reflections preserve angle measures, so the measure of the reflected angle $r(\angle BAD)$ equals the measure of ∠CAD; therefore, $r(\overrightarrow{AB}) = \overrightarrow{AC}$. Reflections also preserve lengths of segments; therefore, the reflection of $\overline{AB}$ still has the same length as $\overline{AB}$. By hypothesis, AB = AC, so the length of the reflection is also equal to AC. Then $r(B) = C$. Using similar reasoning, we can show that $r(C) = B$.

Reflections map rays to rays, so $r(\overrightarrow{BA}) = \overrightarrow{CA}$ and $r(\overrightarrow{BC}) = \overrightarrow{CB}$. Again, since reflections preserve angle measures, the measure of $r(\angle ABC)$ is equal to the measure of ∠ACB.

We conclude that m∠B = m∠C. Equivalently, we can state that ∠B ≅ ∠C. In proofs, we can state that "base angles of an isosceles triangle are equal in measure" or that "base angles of an isosceles triangle are congruent."

Prove Base Angles of an Isosceles are Congruent: SAS

Given: Isosceles △ ABC, with AB = AC

Prove: ∠B ≅ ∠C

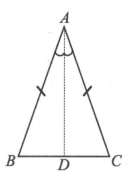

Construction: Draw the angle bisector $\overrightarrow{AD}$ of ∠A, where D is the intersection of the bisector and $\overline{BC}$. We are going to use this auxiliary line towards our SAS criteria.

Exercises

1. Given: $JK = JL$; $\overline{JR}$ bisects $\overline{KL}$
   Prove: $\overline{JR} \perp \overline{KL}$

   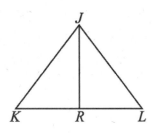

2. Given: $AB = AC$, $XB = XC$
   Prove: $\overline{AX}$ bisects $\angle BAC$

   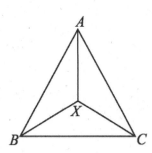

3. Given: $JX = JY$, $KX = LY$
   Prove: $\triangle JKL$ is isosceles

   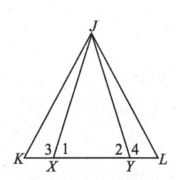

4. Given: △ABC, with m∠CBA = m∠BCA
   Prove: BA = CA
   (Converse of base angles of isosceles triangle)
   Hint: Use a transformation.

5. Given: △ABC, with $\overline{XY}$ is the angle bisector of ∠BYA, and $\overline{BC} \parallel \overline{XY}$
   Prove: YB = YC

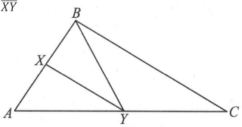

Problem Set

1. Given: $AB = BC$, $AD = DC$
   Prove: $\triangle ADB$ and $\triangle CDB$ are right triangles

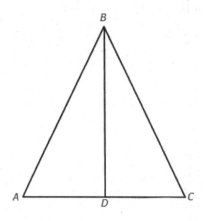

2. Given: $AC = AE$ and $\overline{BF} \parallel \overline{CE}$
   Prove: $AB = AF$

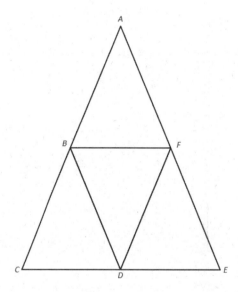

3. In the diagram, $\triangle ABC$ is isosceles with $\overline{AC} \cong \overline{AB}$. In your own words, describe how transformations and the properties of rigid motions can be used to show that $\angle C \cong \angle B$.

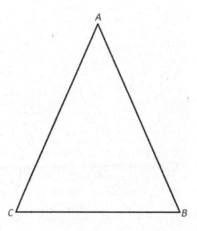

This page intentionally left blank

Lesson 24: Congruence Criteria for Triangles—ASA and SSS

Classwork

Opening Exercise

Use the provided 30° angle as one base angle of an isosceles triangle. Use a compass and straight edge to construct an appropriate isosceles triangle around it.

Compare your constructed isosceles triangle with a neighbor's. Does using a given angle measure guarantee that all the triangles constructed in class have corresponding sides of equal lengths?

Discussion

Today we are going to examine two more triangle congruence criteria, Angle-Side-Angle (ASA) and Side-Side-Side (SSS), to add to the SAS criteria we have already learned. We begin with the ASA criteria.

ANGLE-SIDE-ANGLE TRIANGLE CONGRUENCE CRITERIA (ASA): Given two triangles $\triangle ABC$ and $\triangle A'B'C'$, if $m\angle CAB = m\angle C'A'B'$ (Angle), $AB = A'B'$ (Side), and $m\angle CBA = m\angle C'B'A'$ (Angle), then the triangles are congruent.

PROOF:

We do not begin at the very beginning of this proof. Revisit your notes on the SAS proof, and recall that there are three cases to consider when comparing two triangles. In the most general case, when comparing two distinct triangles, we translate one vertex to another (choose congruent corresponding angles). A rotation brings congruent, corresponding sides together. Since the ASA criteria allows for these steps, we begin here.

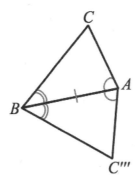

In order to map $\triangle ABC'''$ to $\triangle ABC$, we apply a reflection r across the line AB. A reflection maps A to A and B to B, since they are on line AB. However, we say that $r(C''') = C^*$. Though we know that $r(C''')$ is now in the same half-plane of line AB as C, we cannot assume that C''' maps to C. So we have $r(\triangle ABC''') = \triangle ABC^*$. To prove the theorem, we need to verify that C^* is C.

By hypothesis, we know that $\angle CAB \cong \angle C'''AB$ (recall that $\angle C'''AB$ is the result of two rigid motions of $\angle C'A'B'$, so must have the same angle measure as $\angle C'A'B'$). Similarly, $\angle CBA \cong \angle C'''BA$. Since $\angle CAB \cong r(\angle C'''AB) \cong \angle C^*AB$, and C and C^* are in the same half-plane of line AB, we conclude that $\overrightarrow{AC}$ and $\overrightarrow{AC^*}$ must actually be the same ray. Because the points A and C^* define the same ray as $\overrightarrow{AC}$, the point C^* must be a point somewhere on $\overrightarrow{AC}$. Using the second equality of angles, $\angle CBA \cong r(\angle C'''BA) \cong \angle C^*BA$, we can also conclude that $\overrightarrow{BC}$ and $\overrightarrow{BC^*}$ must be the same ray. Therefore, the point C^* must also be on $\overrightarrow{BC}$. Since C^* is on both $\overrightarrow{AC}$ and $\overrightarrow{BC}$, and the two rays only have one point in common, namely C, we conclude that $C = C^*$.

We have now used a series of rigid motions to map two triangles onto one another that meet the ASA criteria.

SIDE-SIDE-SIDE TRIANGLE CONGRUENCE CRITERIA (SSS): Given two triangles $\triangle ABC$ and $\triangle A'B'C'$, if $AB = A'B'$ (Side), $AC = A'C'$ (Side), and $BC = B'C'$ (Side), then the triangles are congruent.

PROOF:

Again, we do not need to start at the beginning of this proof, but assume there is a congruence that brings a pair of corresponding sides together, namely the longest side of each triangle.

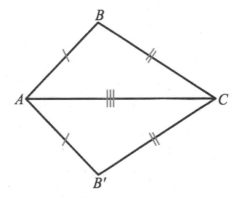

Without any information about the angles of the triangles, we cannot perform a reflection as we have in the proofs for SAS and ASA. What can we do? First we add a construction: Draw an auxiliary line from B to B', and label the angles created by the auxiliary line as r, s, t, and u.

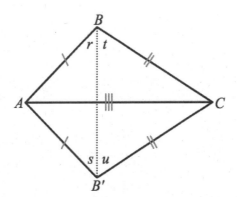

Since $AB = AB'$ and $CB = CB'$, $\triangle ABB'$ and $\triangle CBB'$ are both isosceles triangles respectively by definition. Therefore, $r = s$ because they are base angles of an isosceles triangle ABB'. Similarly, $m\angle t = m\angle u$ because they are base angles of $\triangle CBB'$. Hence, $\angle ABC = m\angle r + m\angle t = m\angle s + m\angle u = m\angle AB'C$. Since $m\angle ABC = m\angle AB'C$, we say that $\triangle ABC \cong \triangle AB'C$ by SAS.

We have now used a series of rigid motions and a construction to map two triangles that meet the SSS criteria onto one another. Note that when using the Side-Side-Side triangle congruence criteria as a reason in a proof, you need only state the congruence and *SSS*. Similarly, when using the Angle-Side-Angle congruence criteria in a proof, you need only state the congruence and *ASA*.

Now we have three triangle congruence criteria at our disposal: SAS, ASA, and SSS. We use these criteria to determine whether or not pairs of triangles are congruent.

Lesson 24: Congruence Criteria for Triangles—ASA and SSS

Exercises

Based on the information provided, determine whether a congruence exists between triangles. If a congruence exists between triangles or if multiple congruencies exist, state the congruencies and the criteria used to determine them.

1. Given: M is the midpoint of $\overline{HP}$, $m\angle H = m\angle P$

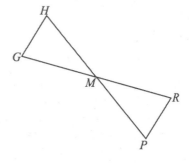

2. Given: Rectangle $JKLM$ with diagonal $\overline{KM}$

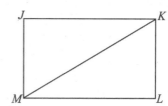

3. Given: $RY = RB$, $AR = XR$

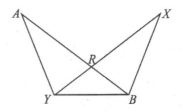

Lesson 24: Congruence Criteria for Triangles—ASA and SSS

4. Given: $m\angle A = m\angle D$, $AE = DE$

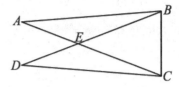

5. Given: $AB = AC$, $BD = \frac{1}{4}AB$, $CE = \frac{1}{4}AC$

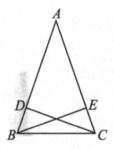

Problem Set

Use your knowledge of triangle congruence criteria to write proofs for each of the following problems.

1. Given: Circles with centers A and B intersect at C and D
   Prove: $\angle CAB \cong \angle DAB$

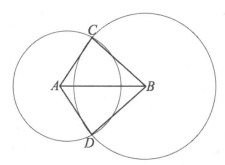

2. Given: $\angle J \cong \angle M, JA = MB, JK = KL = LM$
   Prove: $\overline{KR} \cong \overline{LR}$

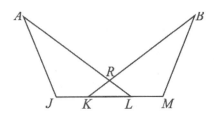

3. Given: $m\angle w = m\angle x$ and $m\angle y = m\angle z$
   Prove: (1) $\triangle ABE \cong \triangle ACE$
   (2) $AB = AC$ and $\overline{AD} \perp \overline{BC}$

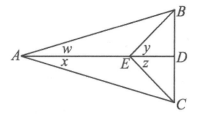

4. After completing the last exercise, Jeanne said, "We also could have been given that $\angle w \cong \angle x$ and $\angle y \cong \angle z$. This would also have allowed us to prove that $\triangle ABE \cong \triangle ACE$." Do you agree? Why or why not?

Lesson 25: Congruence Criteria for Triangles—AAS and HL

Classwork

Opening Exercise

Write a proof for the following question. Once done, compare your proof with a neighbor's.

Given: $DE = DG, EF = GF$

Prove: DF is the angle bisector of $\angle EDG$

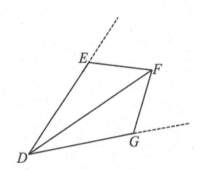

Exploratory Challenge

Today we are going to examine three possible triangle congruence criteria, Angle-Angle-Side (AAS), Side-Side-Angle (SSA), and Angle-Angle-Angle (AAA). Ultimately, only one of the three possible criteria ensures congruence.

ANGLE-ANGLE-SIDE TRIANGLE CONGRUENCE CRITERIA (AAS): Given two triangles $\triangle ABC$ and $\triangle A'B'C'$. If $AB = A'B'$ (Side), $m\angle B = m\angle B'$ (Angle), and $m\angle C = m\angle C'$ (Angle), then the triangles are congruent.

PROOF:

Consider a pair of triangles that meet the AAS criteria. If you knew that two angles of one triangle corresponded to and were equal in measure to two angles of the other triangle, what conclusions can you draw about the third angle of each triangle?

Since the first two angles are equal in measure, the third angles must also be equal in measure.

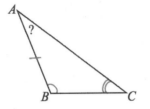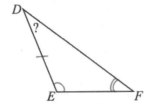

Given this conclusion, which formerly learned triangle congruence criteria can we use to determine if the pair of triangles are congruent?

Therefore, the AAS criterion is actually an extension of the _____ triangle congruence criterion.

Note that when using the Angle-Angle-Side triangle congruence criteria as a reason in a proof, you need only state the congruence and *AAS*.

HYPOTENUSE-LEG TRIANGLE CONGRUENCE CRITERIA (HL): Given two right triangles $\triangle ABC$ and $\triangle A'B'C'$ with right angles B and B'. If $AB = A'B'$ (Leg) and $AC = A'C'$ (Hypotenuse), then the triangles are congruent.

PROOF:

As with some of our other proofs, we do not start at the very beginning, but imagine that a congruence exists so that triangles have been brought together such that $A = A'$ and $C = C'$; the hypotenuse acts as a common side to the transformed triangles.

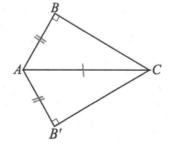

Similar to the proof for SSS, we add a construction and draw $\overline{BB'}$.

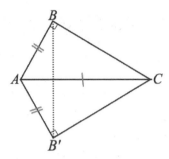

△ ABB' is isosceles by definition, and we can conclude that base angles $m\angle ABB' = m\angle AB'B$. Since $\angle CBB'$ and $\angle CB'B$ are both the complements of equal angle measures ($\angle ABB'$ and $\angle AB'B$), they too are equal in measure. Furthermore, since $m\angle CBB' = m\angle CB'B$, the sides of △ CBB' opposite them are equal in measure: $BC = B'C'$.

Then, by SSS, we can conclude △ $ABC \cong$ △ $A'B'C'$. Note that when using the Hypotenuse-Leg triangle congruence criteria as a reason in a proof, you need only to state the congruence and *HL*.

Criteria that do not determine two triangles as congruent: SSA and AAA

SIDE-SIDE-ANGLE (SSA): Observe the diagrams below. Each triangle has a set of adjacent sides of measures 11 and 9, as well as the non-included angle of 23°. Yet, the triangles are not congruent.

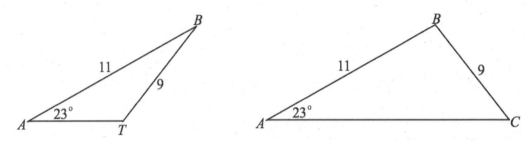

Examine the composite made of both triangles. The sides of length 9 each have been dashed to show their possible locations.

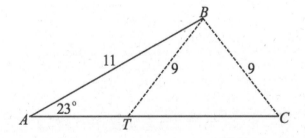

The triangles that satisfy the conditions of SSA cannot *guarantee* congruence criteria. In other words, two triangles under SSA criteria may or may not be congruent; therefore, we cannot categorize SSA as congruence criterion.

A STORY OF FUNCTIONS   Lesson 25 M1

                                                                                        GEOMETRY

ANGLE-ANGLE-ANGLE (AAA): A correspondence exists between △ ABC and △ DEF. Trace △ ABC onto patty paper, and line up corresponding vertices.

Based on your observations, why isn't AAA categorizes as congruence criteria? Is there any situation in which AAA does guarantee congruence?

Even though the angle measures may be the same, the sides can be proportionally larger; you can have similar triangles in addition to a congruent triangle.

List all the triangle congruence criteria here: _____

List the criteria that do not determine congruence here: _____

Examples

1. Given: $\overline{BC} \perp \overline{CD}, \overline{AB} \perp \overline{AD}, m\angle 1 = m\angle 2$
   Prove: △ $BCD \cong$ △ BAD

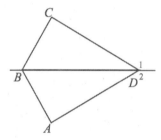

2. Given: $\overline{AD} \perp \overline{BD}, \overline{BD} \perp \overline{BC}, AB = CD$
   Prove: △ $ABD \cong$ △ CDB

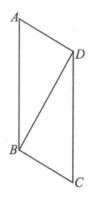

S.158   Lesson 25:   Congruence Criteria for Triangles—AAS and HL

A STORY OF FUNCTIONS
Lesson 25 M1
GEOMETRY

Problem Set

Use your knowledge of triangle congruence criteria to write proofs for each of the following problems.

1. Given: $\overline{AB} \perp \overline{BC}, \overline{DE} \perp \overline{EF}, \overline{BC} \parallel \overline{EF}, AF = DC$
   Prove: $\triangle ABC \cong \triangle DEF$

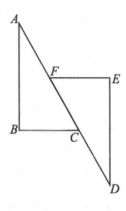

2. In the figure, $\overleftrightarrow{PA} \perp \overleftrightarrow{AR}$ and $\overleftrightarrow{PB} \perp \overleftrightarrow{RB}$ and R is equidistant from $\overleftrightarrow{PA}$ and $\overleftrightarrow{PB}$. Prove that $\overline{PR}$ bisects $\angle APB$.

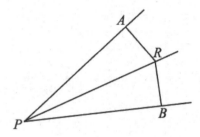

3. Given: $\angle A \cong \angle P, \angle B \cong \angle R, W$ is the midpoint of $\overline{AP}$
   Prove: $\overline{RW} \cong \overline{BW}$

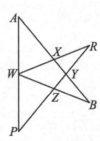

4. Given: $BR = CU$, rectangle $RSTU$
   Prove: $\triangle ARU$ is isosceles

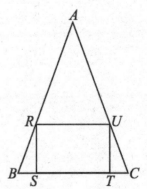

Lesson 25: Congruence Criteria for Triangles—AAS and HL
S.159

This page intentionally left blank

Lesson 26: Triangle Congruency Proofs

Classwork

Exercises

1. Given: $\overline{AB} \perp \overline{BC}, \overline{BC} \perp \overline{DC}$
   $\overline{DB}$ bisects $\angle ABC$, $\overline{AC}$ bisects $\angle DCB$
   $EB = EC$
   Prove: $\triangle BEA \cong \triangle CED$

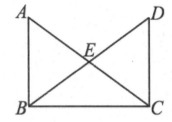

2. Given: $\overline{BF} \perp \overline{AC}$, $\overline{CE} \perp \overline{AB}$
   $AE = AF$
   Prove: $\triangle ACE \cong \triangle ABF$

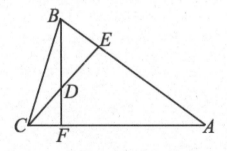

3. Given: $XJ = YK, PX = PY, m\angle ZXJ = m\angle ZYK$
   Prove: $JY = KX$

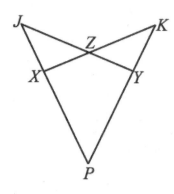

4. Given: $JK = JL, \overline{JK} \parallel \overline{XY}$
   Prove: $XY = XL$

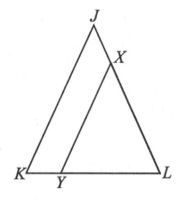

5. Given: ∠1 ≅ ∠2, ∠3 ≅ ∠4
   Prove: $\overline{AC} \cong \overline{BD}$

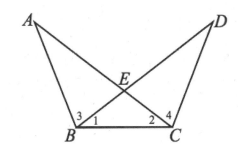

6. Given: $m\angle 1 = m\angle 2$, $m\angle 3 = m\angle 4$, $AB = AC$
   Prove: (a) △ ABD ≅ △ ACD
   (b) $m\angle 5 = m\angle 6$

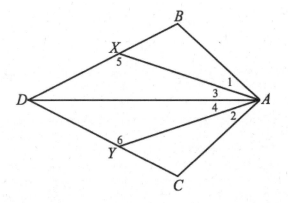

Lesson 26: Triangle Congruency Proofs

Problem Set

Use your knowledge of triangle congruence criteria to write a proof for the following:

In the figure, $\overline{RX}$ and $\overline{RY}$ are the perpendicular bisectors of $\overline{AB}$ and $\overline{AC}$, respectively.

Prove: (a) $\triangle RAX \cong \triangle RBX$

(b) $\overline{RA} \cong \overline{RB} \cong \overline{RC}$

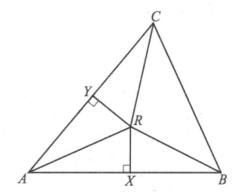

Lesson 27: Triangle Congruency Proofs

Classwork

Exercises

1. Given: $AB = AC, RB = RC$
   Prove: $SB = SC$

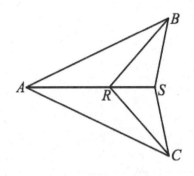

2. Given: Square $ABCS \cong$ Square $EFGS$,
   $\overleftrightarrow{RAB}, \overleftrightarrow{REF}$
   Prove: $\triangle ASR \cong \triangle ESR$

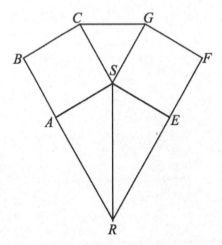

3. Given: $JK = JL, JX = JY$
   Prove: $KX = LY$

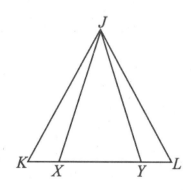

4. Given: $\overline{AD} \perp \overline{DR}, \overline{AB} \perp \overline{BR},$
   $\overline{AD} \cong \overline{AB}$
   Prove: $\angle DCR \cong \angle BCR$

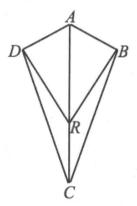

5. Given: $AR = AS, BR = CS,$
   $\overline{RX} \perp \overline{AB}, \overline{SY} \perp \overline{AC}$
   Prove: $BX = CY$

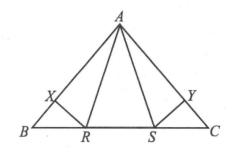

6. Given: $AX = BX, m\angle AMB = m\angle AYZ = 90°$
   Prove: $NY = NM$

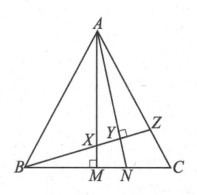

Problem Set

Use your knowledge of triangle congruence criteria to write a proof for the following:

In the figure $\overline{BE} \cong \overline{CE}$, $\overline{DC} \perp \overline{AB}$, and $\overline{BE} \perp \overline{AC}$; prove $\overline{AE} \cong \overline{RE}$.

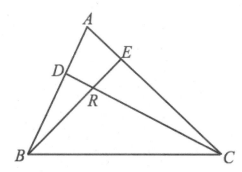

Lesson 28: Properties of Parallelograms

Classwork

Opening Exercise

a. If the triangles are congruent, state the congruence.

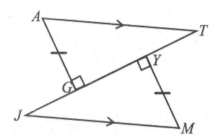

b. Which triangle congruence criterion guarantees part 1?

c. $\overline{TG}$ corresponds with

Discussion

How can we use our knowledge of triangle congruence criteria to establish other geometry facts? For instance, what can we now prove about the properties of parallelograms?

To date, we have defined a parallelogram to be a quadrilateral in which both pairs of opposite sides are parallel. However, we have assumed other details about parallelograms to be true, too. We assume that:

- Opposite sides are congruent.
- Opposite angles are congruent.
- Diagonals bisect each other.

Let us examine why each of these properties is true.

Example 1

If a quadrilateral is a parallelogram, then its opposite sides and angles are equal in measure. Complete the diagram, and develop an appropriate *Given* and *Prove* for this case. Use triangle congruence criteria to demonstrate why opposite sides and angles of a parallelogram are congruent.

Given: _____

Prove: _____

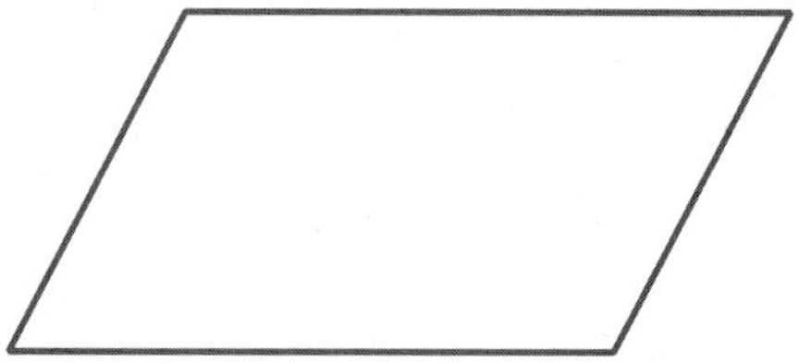

Construction: Label the quadrilateral $ABCD$, and mark opposite sides as parallel. Draw diagonal $\overline{BD}$.

Example 2

If a quadrilateral is a parallelogram, then the diagonals bisect each other. Complete the diagram, and develop an appropriate *Given* and *Prove* for this case. Use triangle congruence criteria to demonstrate why diagonals of a parallelogram bisect each other. Remember, now that we have proved opposite sides and angles of a parallelogram to be congruent, we are free to use these facts as needed (i.e., $AD = CB$, $AB = CD$, $\angle A \cong \angle C$, $\angle B \cong \angle D$).

Given: _____

Prove: _____

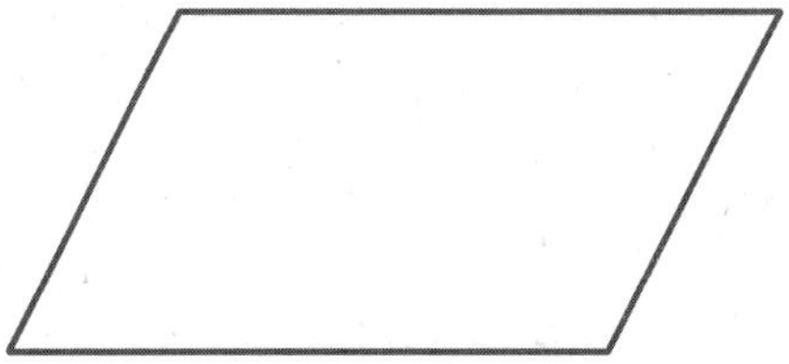

Construction: Label the quadrilateral $ABCD$. Mark opposite sides as parallel. Draw diagonals $\overline{AC}$ and $\overline{BD}$.

Now we have established why the properties of parallelograms that we have assumed to be true are in fact true. By extension, these facts hold for any type of parallelogram, including rectangles, squares, and rhombuses. Let us look at one last fact concerning rectangles. We established that the diagonals of general parallelograms bisect each other. Let us now demonstrate that a rectangle has congruent diagonals.

Example 3

If the parallelogram is a rectangle, then the diagonals are equal in length. Complete the diagram, and develop an appropriate *Given* and *Prove* for this case. Use triangle congruence criteria to demonstrate why diagonals of a rectangle are congruent. As in the last proof, remember to use any already proven facts as needed.

Given: _____

Prove: _____

Construction: Label the rectangle $GHIJ$. Mark opposite sides as parallel, and add small squares at the vertices to indicate 90° angles. Draw diagonals $\overline{GI}$ and $\overline{HJ}$.

Converse Properties: Now we examine the converse of each of the properties we proved. Begin with the property, and prove that the quadrilateral is in fact a parallelogram.

Example 4

If both pairs of opposite angles of a quadrilateral are equal, then the quadrilateral is a parallelogram. Draw an appropriate diagram, and provide the relevant *Given* and *Prove* for this case.

Given: _____

Prove: _____

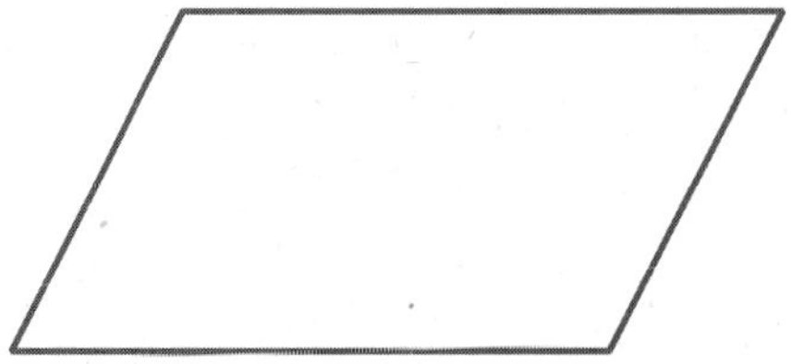

Construction: Label the quadrilateral $ABCD$. Mark opposite angles as congruent. Draw diagonal $\overline{BD}$. Label the measures of $\angle A$ and $\angle C$ as $x°$. Label the measures of the four angles created by $\overline{BD}$ as $r°$, $s°$, $t°$, and $u°$.

Lesson 28: Properties of Parallelograms

A STORY OF FUNCTIONS   **Lesson 28** **M1**

                                                                                              GEOMETRY

Example 5

If the opposite sides of a quadrilateral are equal, then the quadrilateral is a parallelogram. Draw an appropriate diagram, and provide the relevant *Given* and *Prove* for this case.

Given: _____

Prove: _____

Construction: Label the quadrilateral $ABCD$, and mark opposite sides as equal. Draw diagonal $\overline{BD}$.

Example 6

If the diagonals of a quadrilateral bisect each other, then the quadrilateral is a parallelogram. Draw an appropriate diagram, and provide the relevant *Given* and *Prove* for this case. Use triangle congruence criteria to demonstrate why the quadrilateral is a parallelogram.

Given: _____

Prove: _____

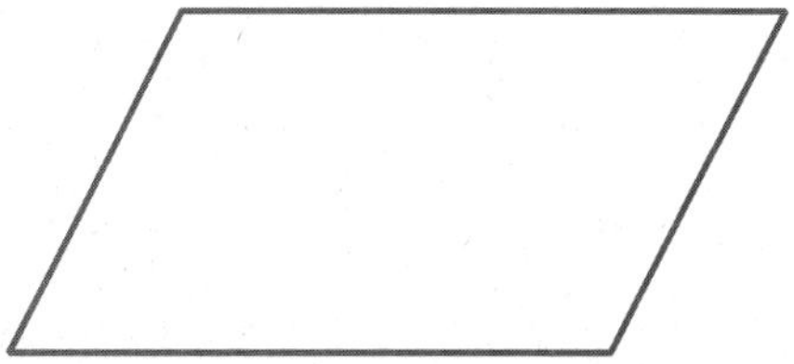

Construction: Label the quadrilateral $ABCD$, and mark opposite sides as equal. Draw diagonals $\overline{AC}$ and $\overline{BD}$.

Example 7

If the diagonals of a parallelogram are equal in length, then the parallelogram is a rectangle. Complete the diagram, and develop an appropriate *Given* and *Prove* for this case.

Given: _____

Prove: _____

Construction: Label the quadrilateral $GHIJ$. Draw diagonals $\overline{GI}$ and $\overline{HJ}$.

A STORY OF FUNCTIONS
Lesson 28 M1

GEOMETRY

Problem Set

Use the facts you have established to complete exercises involving different types of parallelograms.

1. Given: $\overline{AB} \parallel \overline{CD}$, $AD = AB$, $CD = CB$
   Prove: $ABCD$ is a rhombus.

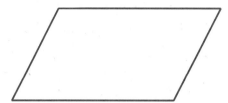

2. Given: Rectangle $RSTU$, M is the midpoint of $\overline{RS}$.
   Prove: $\triangle UMT$ is isosceles.

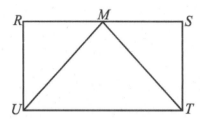

3. Given: $ABCD$ is a parallelogram, $\overline{RD}$ bisects $\angle ADC$, $\overline{SB}$ bisects $\angle CBA$.
   Prove: $DRBS$ is a parallelogram.

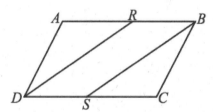

4. Given: $DEFG$ is a rectangle, $WE = YG$, $WX = YZ$
   Prove: $WXYZ$ is a parallelogram.

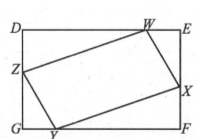

5. Given: Parallelogram $ABFE$, $CR = DS$, $\overline{ABC}$ and $\overline{DEF}$ are segments.
   Prove: $BR = SE$

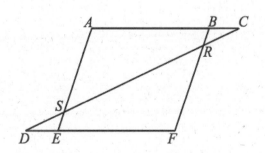

Lesson 28: Properties of Parallelograms

This page intentionally left blank

Lesson 29: Special Lines in Triangles

Classwork

Opening Exercise

Construct the midsegment of the triangle below. A midsegment is a line segment that joins the midpoints of two sides of a triangle or trapezoid. For the moment, we will work with a triangle.

    a.   Use your compass and straightedge to determine the midpoints of $\overline{AB}$ and $\overline{AC}$ as X and Y, respectively.

    b.   Draw midsegment $\overline{XY}$.

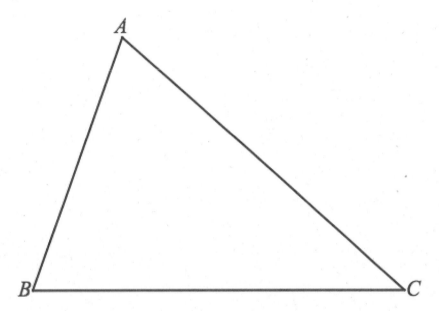

Compare $\angle AXY$ and $\angle ABC$; compare $\angle AYX$ and $\angle ACB$. Without using a protractor, what would you guess is the relationship between these two pairs of angles? What are the implications of this relationship?

Discussion

Note that though we chose to determine the midsegment of $\overline{AB}$ and $\overline{AC}$, we could have chosen any two sides to work with. Let us now focus on the properties associated with a midsegment.

The midsegment of a triangle is parallel to the third side of the triangle and half the length of the third side of the triangle.

We can prove these properties to be true. Continue to work with the figure from the Opening Exercise.

Given: $\overline{XY}$ is a midsegment of $\triangle ABC$.

Prove: $\overline{XY} \parallel \overline{BC}$ and $XY = \frac{1}{2}BC$

Construct the following: In the Opening Exercise figure, draw $\triangle YGC$ according to the following steps. Extend $\overline{XY}$ to point G so that $YG = XY$. Draw $\overline{GC}$.

(1) What is the relationship between XY and YG? Explain why. _____

(2) What is the relationship between $\angle AYX$ and $\angle GYC$? Explain why. _____

(3) What is the relationship between AY and YC? Explain why. _____

(4) What is the relationship between $\triangle AXY$ and $\triangle CGY$? Explain why. _____

(5) What is the relationship between GC and AX? Explain why. _____

(6) Since $AX = BX$, what other conclusion can be drawn? Explain why. _____

(7) What is the relationship between $m\angle AXY$ and $m\angle YGC$? Explain why. _____

(8) Based on (7), what other conclusion can be drawn about $\overline{AB}$ and $\overline{GC}$? Explain why. _____

_____

(9) What conclusion can be drawn about $BXGC$ based on (7) and (8)? Explain why. _____

_____

(10) Based on (9), what is the relationship between XG and BC? _____

_____

(11) Since $YG = XY$, $XG =$ _____ XY. Explain why. _____

_____

(12) This means $BC =$ _____ XY. Explain why. _____

_____

(13) Or by division, $XY =$ _____ BC.

Note that Steps (9) and (13) demonstrate our *Prove* statement.

Exercises 1–4

Apply what you know about the properties of midsegments to solve the following exercises.

1. $x =$ _____
   Perimeter of $\triangle ABC =$ _____

2. $x =$ _____
   $y =$ _____

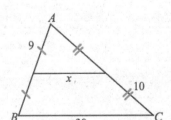

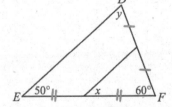

Lesson 29: Special Lines in Triangles

A STORY OF FUNCTIONS   Lesson 29   M1
                                                                          GEOMETRY

3. In △ RST, the midpoints of each side have been marked by points X, Y, and Z.
    - Mark the halves of each side divided by the midpoint with a congruency mark. Remember to distinguish congruency marks for each side.
    - Draw midsegments $\overline{XY}, \overline{YZ}$, and $\overline{XZ}$. Mark each midsegment with the appropriate congruency mark from the sides of the triangle.

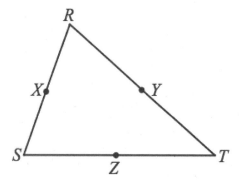

   a. What conclusion can you draw about the four triangles within △ RST? Explain why. _____
   _____

   b. State the appropriate correspondences among the four triangles within △ RST. _____
   _____

   c. State a correspondence between △ RST and any one of the four small triangles. _____
   _____

4. Find x.

    $x =$ _____

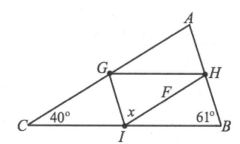

S.182   Lesson 29:   Special Lines in Triangles

Problem Set

Use your knowledge of triangle congruence criteria to write proofs for each of the following problems.

1. $\overline{WX}$ is a midsegment of $\triangle ABC$, and $\overline{YZ}$ is a midsegment of $\triangle CWX$. $BX = AW$

   a. What can you conclude about $\angle A$ and $\angle B$? Explain why.

   b. What is the relationship of the lengths $\overline{YZ}$ and $\overline{AB}$?

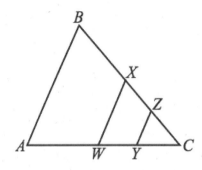

2. W, X, Y, and Z are the midpoints of $\overline{AD}, \overline{AB}, \overline{BC}$, and $\overline{CD}$, respectively. $AD = 18$, $WZ = 11$, and $BX = 5$. $m\angle WAC = 33°$, $m\angle RYX = 74°$

   a. $m\angle DZW = $ _____

   b. Perimeter of $ABYW = $ _____

   c. Perimeter of $ABCD = $ _____

   d. $m\angle WAX = $ _____

   $m\angle B = $ _____

   $m\angle YCZ = $ _____

   $m\angle D = $ _____

   e. What kind of quadrilateral is $ABCD$?

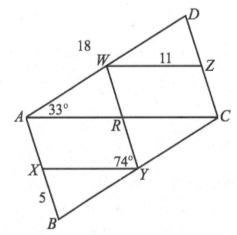

This page intentionally left blank

Lesson 30: Special Lines in Triangles

Classwork

Opening Exercise

In △ ABC to the right, D is the midpoint of $\overline{AB}$, E is the midpoint of $\overline{BC}$, and F is the midpoint of $\overline{AC}$. Complete each statement below.

$\overline{DE}$ is parallel to _____ and measures _____ the length of _____.

$\overline{DF}$ is parallel to _____ and measures _____ the length of _____.

$\overline{EF}$ is parallel to _____ and measures _____ the length of _____.

Discussion

In the previous two lessons, we proved that (a) the midsegment of a triangle is parallel to the third side and half the length of the third side and (b) diagonals of a parallelogram bisect each other. We use both of these facts to prove the following assertion:

All medians of a triangle are _____. That is, the three medians of a triangle (the segments connecting each vertex to the midpoint of the opposite side) meet at a single point. This point of concurrency is called the _____, or the center of gravity, of the triangle. The proof also shows a length relationship for each median: The length from the vertex to the centroid is _____ the length from the centroid to the midpoint of the side.

Example 1

Provide a valid reason for each step in the proof below.

Given: △ABC with D, E, and F the midpoints of sides $\overline{AB}$, $\overline{BC}$, and $\overline{AC}$, respectively

Prove: The three medians of △ABC meet at a single point.

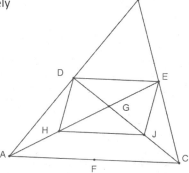

(1) Draw midsegment $\overline{DE}$. Draw $\overline{AE}$ and $\overline{DC}$; label their intersection as point G.

(2) Construct and label the midpoint of $\overline{AG}$ as point H and the midpoint of $\overline{GC}$ as point J.

(3) $\overline{DE} \parallel \overline{AC}$,

_____

(4) $\overline{HJ} \parallel \overline{AC}$,

_____

(5) $\overline{DE} \parallel \overline{HJ}$,

_____

(6) $DE = \frac{1}{2}AC$ and $HJ = \frac{1}{2}AC$,

_____

(7) DEJH is a parallelogram.

_____

(8) $HG = EG$ and $JG = DG$,

_____

(9) $AH = HG$ and $CJ = JG$,

_____

(10) $AH = HG = GE$ and $CJ = JG = GD$,

_____

(11) $AG = 2GE$ and $CG = 2GD$,

_____

(12) We can complete Steps (1)–(11) to include the median from B; the third median, $\overline{BF}$, passes through point G, which divides it into two segments such that the longer part is twice the shorter.

(13) The intersection point of the medians divides each median into two parts with lengths in a ratio of 2:1; therefore, all medians are concurrent at that point.

The three medians of a triangle are concurrent at the _____, or the center of gravity. This point of concurrency divides the length of each median in a ratio of _____; the length from the vertex to the centroid is _____ the length from the centroid to the midpoint of the side.

Example 2

In $\triangle ABC$, the medians are concurrent at F. $DF = 4$, $BF = 16$, $AG = 30$. Find each of the following measures.

a. $FC = $ _____

b. $DC = $ _____

c. $AF = $ _____

d. $BE = $ _____

e. $FG = $ _____

f. $EF = $ _____

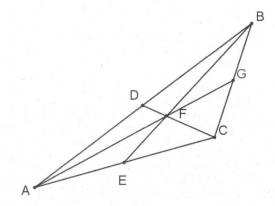

Lesson 30: Special Lines in Triangles

A STORY OF FUNCTIONS   Lesson 30 M1
                                                                                         GEOMETRY

Example 3

In the figure to the right, $\triangle ABC$ is reflected over $\overline{AB}$ to create $\triangle ABD$. Points P, E, and F are midpoints of $\overline{AB}$, $\overline{BD}$, and $\overline{BC}$, respectively. If $AH = AG$, prove that $PH = GP$.

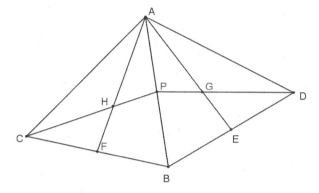

Lesson 30:   Special Lines in Triangles

Problem Set

Ty is building a model of a hang glider using the template below. To place his supports accurately, Ty needs to locate the center of gravity on his model.

1. Use your compass and straightedge to locate the center of gravity on Ty's model.

2. Explain what the center of gravity represents on Ty's model.

3. Describe the relationship between the longer and shorter sections of the line segments you drew as you located the center of gravity.

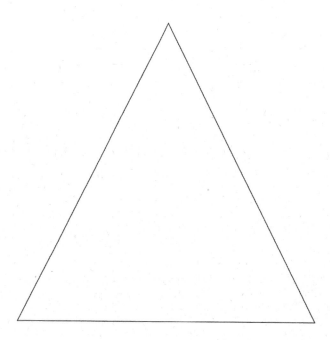

Lesson 30: Special Lines in Triangles

This page intentionally left blank

Lesson 31: Construct a Square and a Nine-Point Circle

Classwork

Opening Exercise

With a partner, use your construction tools and what you learned in Lessons 1–5 to attempt the construction of a square. Once you are satisfied with your construction, write the instructions to perform the construction.

Exploratory Challenge

Now, we are going to construct a nine-point circle. What is meant by the phrase *nine-point circle*?

_____

_____

Steps to construct a nine-point circle:

1. Draw a triangle ABC.
2. Construct the midpoints of the sides $\overline{AB}$, $\overline{BC}$, and $\overline{CA}$, and label them as L, M, and N, respectively.
3. Construct the perpendicular from each vertex to the opposite side of the triangle (each is called an *altitude*).
4. Label the intersection of the altitude from C to $\overline{AB}$ as D, the intersection of the altitude from A to $\overline{BC}$ as E, and of the altitude from B to $\overline{CA}$ as F.
5. The altitudes are concurrent at a point; label it H.
6. Construct the midpoints of $\overline{AH}$, $\overline{BH}$, $\overline{CH}$, and label them X, Y, and Z, respectively.
7. The nine points, $L, M, N, D, E, F, X, Y, Z$, are the points that define the nine-point circle.

Lesson 31

GEOMETRY

Example

On a blank white sheet of paper, construct a nine-point circle using a different triangle than you used during the notes. Does the type of triangle you start with affect the construction of the nine-point circle?

Lesson 31: Construct a Square and a Nine-Point Circle

Problem Set

Construct square $ABCD$ and square $GHIJ$ so that

a. Each side of $GHIJ$ is half the length of each $ABCD$.
b. $\overline{AB}$ contains $\overline{GH}$.
c. The midpoint of $\overline{AB}$ is also the midpoint of $\overline{GH}$.

Lesson 32: Construct a Nine-Point Circle

Classwork

Opening Exercise

During this unit we, have learned many constructions. Now that you have mastered these constructions, write a list of advice for someone who is about to learn the constructions you have learned for the first time. What did and did not help you? What tips did you wish you had at the beginning that would have made it easier along the way?

Exploratory Challenge 1

Yesterday, we began the nine-point circle construction. What did we learn about the triangle that we start our construction with? Where did we stop in the construction?

We continue our construction today.

There are two constructions for finding the center of the nine-point circle. With a partner, work through both constructions.

Construction 1

1. To find the center of the circle, draw inscribed △ LMN.
2. Find the circumcenter of △ LMN, and label it as U.

Recall that the circumcenter of a triangle is the center of the circle that circumscribes the triangle, which, in this case, is the nine-point circle.

Construction 2

1. Construct the circle that circumscribes △ ABC.
2. Find the circumcenter of △ ABC, which is the center of the circle that circumscribes △ ABC. Label its center CC.
3. Draw the segment that joins point H (the orthocenter from the construction of the nine-point circle in Lesson 31) to the point CC.
4. Find the midpoint of the segment you drew in Step 3, and label that point U.

Describe the relationship between the midpoint you found in Step 4 of the second construction and the point U in the first construction.

Exploratory Challenge 2

Construct a square $ABCD$. Pick a point E between B and C, and draw a segment from point A to a point E. The segment forms a right triangle and a trapezoid out of the square. Construct a nine-point circle using the right triangle.

Problem Set

Take a blank sheet of $8\frac{1}{2}$ inch by 11 inch white paper, and draw a triangle with vertices on the edge of the paper. Construct a nine-point circle within this triangle. Then, draw a triangle with vertices on that nine-point circle, and construct a nine-point circle within that. Continue constructing nine-point circles until you no longer have room inside your constructions.

This page intentionally left blank

Lesson 33: Review of the Assumptions

Classwork

Review Exercises

We have covered a great deal of material in Module 1. Our study has included definitions, geometric assumptions, geometric facts, constructions, unknown angle problems and proofs, transformations, and proofs that establish properties we previously took for granted.

In the first list below, we compile all of the geometric assumptions we took for granted as part of our reasoning and proof-writing process. Though these assumptions were only highlights in lessons, these assumptions form the basis from which all other facts can be derived (e.g., the other facts presented in the table). College-level geometry courses often do an in-depth study of the assumptions.

The latter tables review the facts associated with problems covered in Module 1. Abbreviations for the facts are within brackets.

Geometric Assumptions (Mathematicians call these *axioms*.)

1. (Line) Given any two distinct points, there is exactly one line that contains them.
2. (Plane Separation) Given a line contained in the plane, the points of the plane that do not lie on the line form two sets, called *half-planes*, such that
    a. Each of the sets is convex.
    b. If P is a point in one of the sets and Q is a point in the other, then $\overline{PQ}$ intersects the line.
3. (Distance) To every pair of points A and B there corresponds a real number dist $(A, B) \geq 0$, called the *distance* from A to B, so that
    a. $\text{dist}(A, B) = \text{dist}(B, A)$
    b. $\text{dist}(A, B) \geq 0$, and $\text{dist}(A, B) = 0 \iff A$ and B coincide.
4. (Ruler) Every line has a coordinate system.
5. (Plane) Every plane contains at least three noncollinear points.
6. (Basic Rigid Motions) Basic rigid motions (e.g., rotations, reflections, and translations) have the following properties:
    a. Any basic rigid motion preserves lines, rays, and segments. That is, for any basic rigid motion of the plane, the image of a line is a line, the image of a ray is a ray, and the image of a segment is a segment.
    b. Any basic rigid motion preserves lengths of segments and angle measures of angles.
7. (180° Protractor) To every $\angle AOB$, there corresponds a real number $m\angle AOB$, called the *degree* or *measure* of the angle, with the following properties:
    a. $0° < m\angle AOB < 180°$
    b. Let $\overrightarrow{OB}$ be a ray on the edge of the half-plane H. For every r such that $0° < r° < 180°$, there is exactly one ray $\overrightarrow{OA}$ with A in H such that $m\angle AOB = r°$.
    c. If C is a point in the interior of $\angle AOB$, then $m\angle AOC + m\angle COB = m\angle AOB$.
    d. If two angles $\angle BAC$ and $\angle CAD$ form a linear pair, then they are supplementary (e.g., $m\angle BAC + m\angle CAD = 180°$).
8. (Parallel Postulate) Through a given external point, there is at most one line parallel to a given line.

Lesson 33: Review of the Assumptions

S.199

Fact/Property	Guiding Questions/Applications	Notes/Solutions
Two angles that form a linear pair are supplementary.	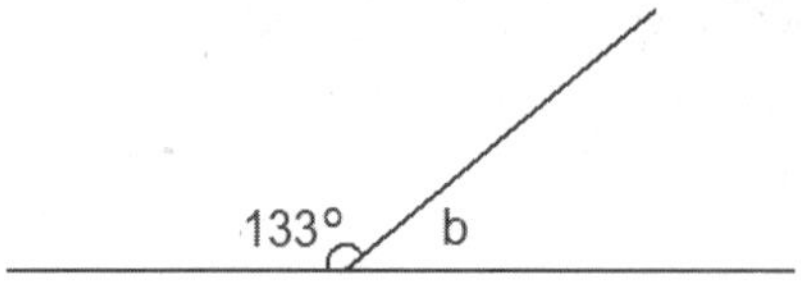 133° b	
The sum of the measures of all adjacent angles formed by three or more rays with the same vertex is 360°.	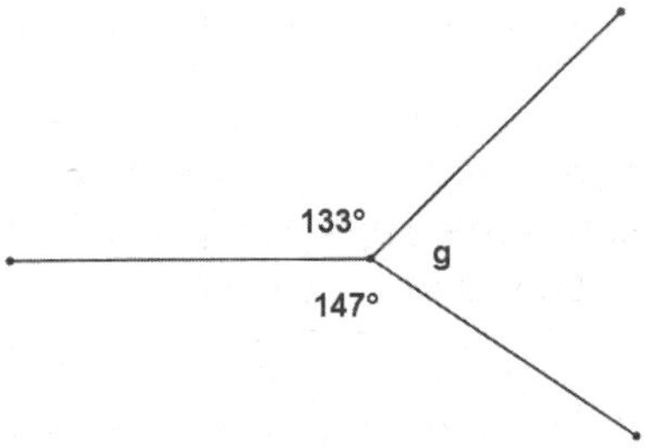 133° g 147°	
Vertical angles have equal measure.	Use the fact that linear pairs form supplementary angles to prove that vertical angles are equal in measure.	
The bisector of an angle is a ray in the interior of the angle such that the two adjacent angles formed by it have equal measure.	In the diagram below, $\overline{BC}$ is the bisector of $\angle ABD$, which measures 64°. What is the measure of $\angle ABC$? 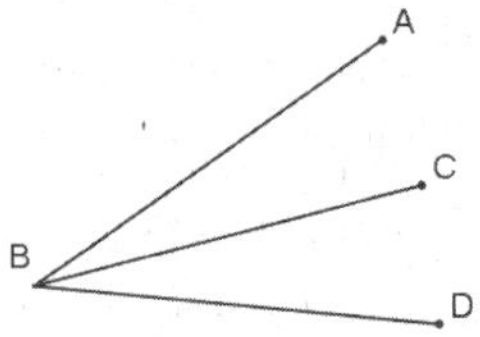	
The perpendicular bisector of a segment is the line that passes through the midpoint of a line segment and is perpendicular to the line segment.	In the diagram below, $\overline{DC}$ is the perpendicular bisector of $\overline{AB}$, and $\overline{CE}$ is the angle bisector of $\angle ACD$. Find the measures of $\overline{AC}$ and $\angle ECD$. 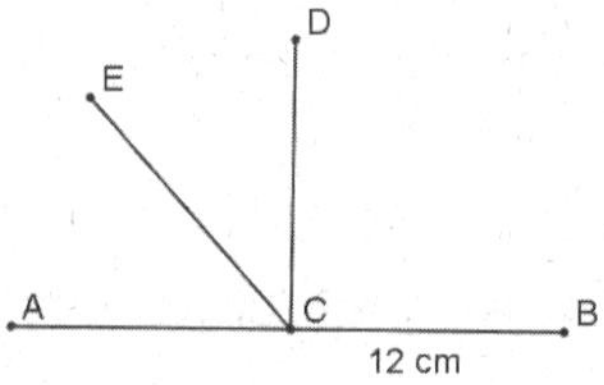 12 cm	

Lesson 33: Review of the Assumptions

The sum of the 3 angle measures of any triangle is 180°.	Given the labeled figure below, find the measures of ∠DEB and ∠ACE. Explain your solutions. 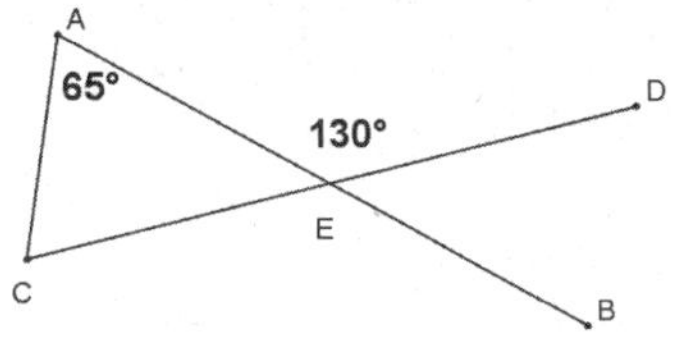	
When one angle of a triangle is a right angle, the sum of the measures of the other two angles is 90°.	This fact follows directly from the preceding one. How is simple arithmetic used to extend the angle sum of a triangle property to justify this property?	
An exterior angle of a triangle is equal to the sum of its two opposite interior angles.	In the diagram below, how is the exterior angle of a triangle property proved? 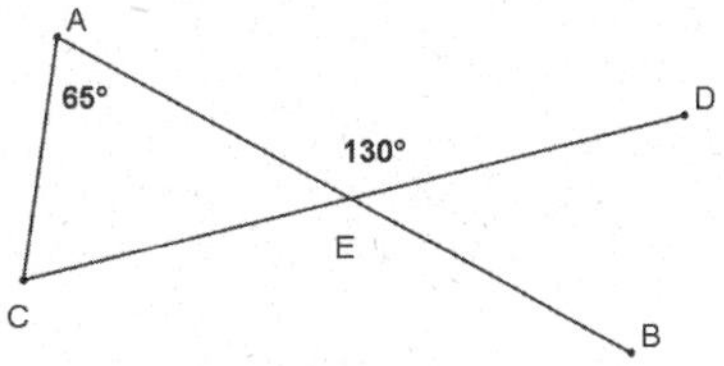	
Base angles of an isosceles triangle are congruent.	The triangle in the figure above is isosceles. How do we know this?	
All angles in an equilateral triangle have equal measure. [equilat. △]	If the figure above is changed slightly, it can be used to demonstrate the equilateral triangle property. Explain how this can be demonstrated. 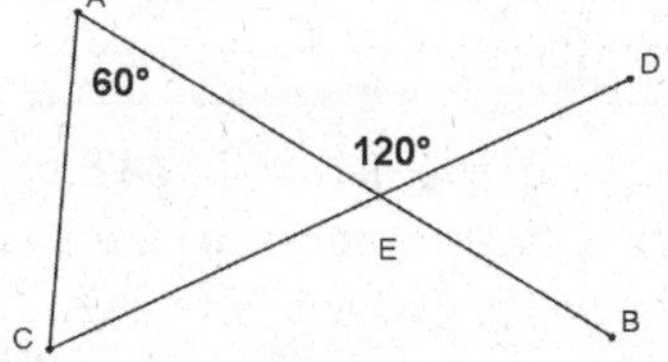	

Lesson 33: Review of the Assumptions

The facts and properties in the immediately preceding table relate to angles and triangles. In the table below, we review facts and properties related to parallel lines and transversals.

Fact/Property	Guiding Questions/Applications	Notes/Solutions
If a transversal intersects two parallel lines, then the measures of the corresponding angles are equal.	Why does the property specify *parallel* lines?	
If a transversal intersects two lines such that the measures of the corresponding angles are equal, then the lines are parallel.	The converse of a statement turns the relevant property into an *if and only if* relationship. Explain how this is related to the guiding question about corresponding angles.	
If a transversal intersects two parallel lines, then the interior angles on the same side of the transversal are supplementary.	This property is proved using (in part) the corresponding angles property. Use the diagram below ($\overline{AB} \parallel \overline{CD}$) to prove that $\angle AGH$ and $\angle CHG$ are supplementary. 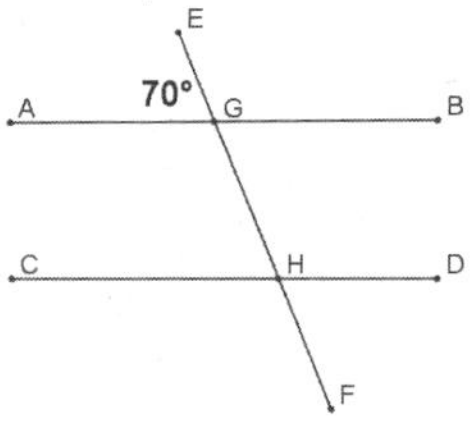	
If a transversal intersects two lines such that the same side interior angles are supplementary, then the lines are parallel.	Given the labeled diagram below, prove that $\overline{AB} \parallel \overline{CD}$. 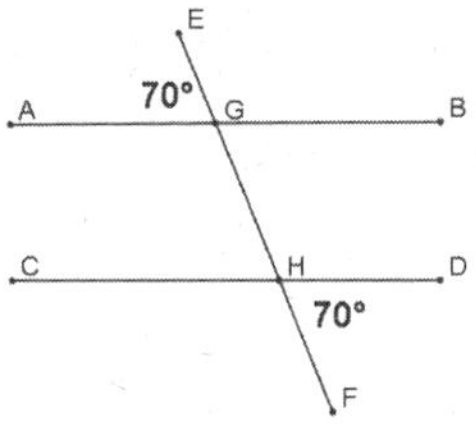	
If a transversal intersects two parallel lines, then the measures of alternate interior angles are equal.	1. Name both pairs of alternate interior angles in the diagram above. 2. How many different angle measures are in the diagram?	
If a transversal intersects two lines such that measures of the alternate interior angles are equal, then the lines are parallel.	Although not specifically stated here, the property also applies to *alternate exterior angles*. Why is this true?	

Lesson 33: Review of the Assumptions

Problem Set

Use any of the assumptions, facts, and/or properties presented in the tables above to find x and y in each figure below. Justify your solutions.

1. $x = $ _____

   $y = $ _____

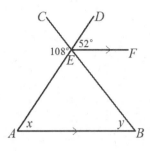

2. You need to draw an auxiliary line to solve this problem.

   $x = $ _____

   $y = $ _____

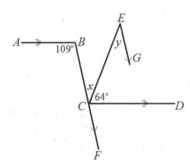

3. $x = $ _____

   $y = $ _____

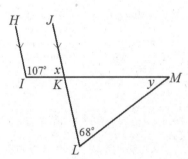

4. Given the labeled diagram at the right, prove that $\angle VWX \cong \angle XYZ$.

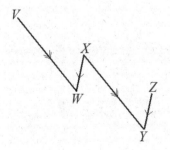

Lesson 33: Review of the Assumptions

Lesson 34: Review of the Assumptions

Classwork

Assumption/Fact/Property	Guiding Questions/Applications	Notes/Solutions
Given two triangles ABC and $A'B'C'$ so that $AB = A'B'$ (Side), $m\angle A = m\angle A'$ (Angle), and $AC = A'C'$ (Side), then the triangles are congruent. [SAS]	The figure below is a parallelogram $ABCD$. What parts of the parallelogram satisfy the SAS triangle congruence criteria for $\triangle ABD$ and $\triangle CDB$? Describe a rigid motion(s) that maps one onto the other. (Consider drawing an auxiliary line.) 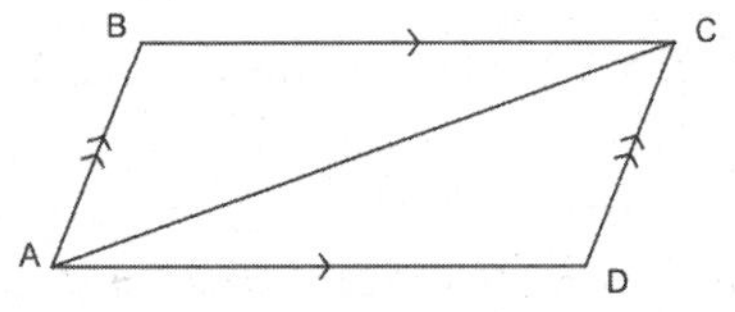	
Given two triangles ABC and $A'B'C'$, if $m\angle A = m\angle A'$ (Angle), $AB = A'B'$ (Side), and $m\angle B = m\angle B'$ (Angle), then the triangles are congruent. [ASA]	In the figure below, $\triangle CDE$ is the image of the reflection of $\triangle ABE$ across line FG. Which parts of the triangle can be used to satisfy the ASA congruence criteria? 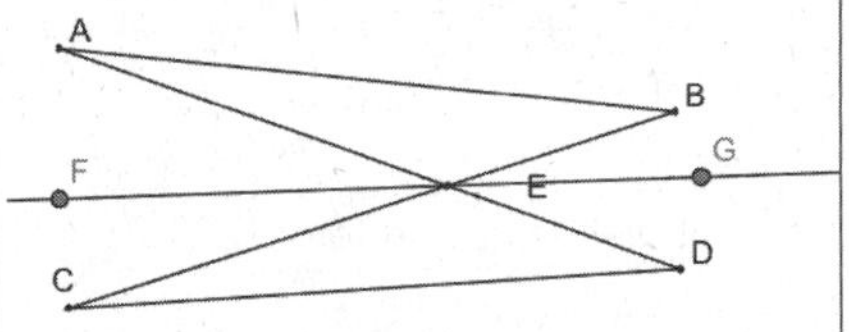	
Given two triangles ABC and $A'B'C'$, if $AB = A'B'$ (Side), $AC = A'C'$ (Side), and $BC = B'C'$ (Side), then the triangles are congruent. [SSS]	$\triangle ABC$ and $\triangle ADC$ are formed from the intersections and center points of circles A and C. Prove $\triangle ABC \cong \triangle ADC$ by SSS. 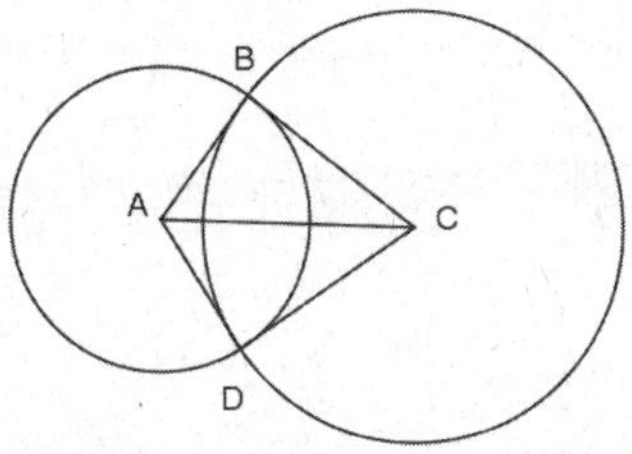	

Given two triangles ABC and $A'B'C'$, if $AB = A'B'$ (Side), $m\angle B = m\angle B'$ (Angle), and $m\angle C = m\angle C'$ (Angle), then the triangles are congruent. [AAS]	The AAS congruence criterion is essentially the same as the ASA criterion for proving triangles congruent. Why is this true? 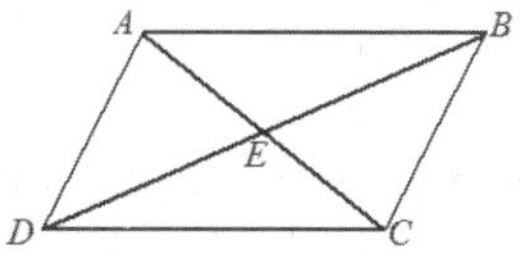	
Given two right triangles ABC and $A'B'C'$ with right angles $\angle B$ and $\angle B'$, if $AB = A'B'$ (Leg) and $AC = A'C'$ (Hypotenuse), then the triangles are congruent. [HL]	In the figure below, CD is the perpendicular bisector of AB, and $\triangle ABC$ is isosceles. Name the two congruent triangles appropriately, and describe the necessary steps for proving them congruent using HL. 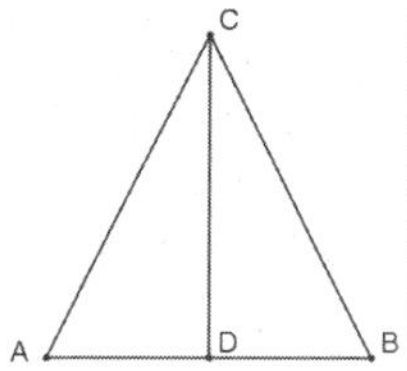	
The opposite sides of a parallelogram are congruent.	In the figure below, $BE \cong DE$ and $\angle CBE \cong \angle ADE$. Prove $ABCD$ is a parallelogram.	
The opposite angles of a parallelogram are congruent.		
The diagonals of a parallelogram bisect each other.		
The midsegment of a triangle is a line segment that connects the midpoints of two sides of a triangle; the midsegment is parallel to the third side of the triangle and is half the length of the third side.	$\overline{DE}$ is the midsegment of $\triangle ABC$. Find the perimeter of $\triangle ABC$, given the labeled segment lengths. 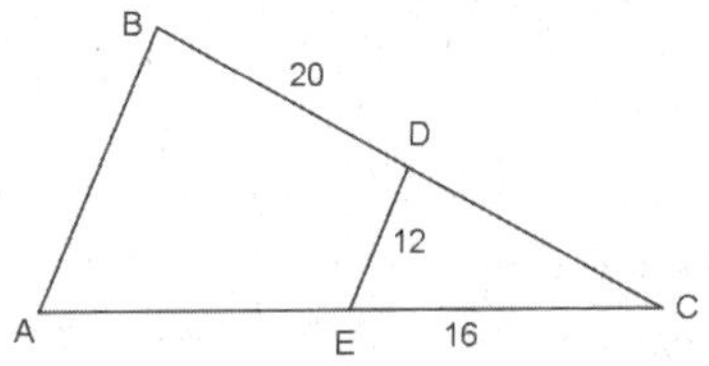	
The three medians of a triangle are concurrent at the centroid; the centroid divides each median into two parts, from vertex to centroid and centroid to midpoint, in a ratio of $2:1$.	If $\overline{AE}$, $\overline{BF}$, and $\overline{CD}$ are medians of $\triangle ABC$, find the length of BG, GE, and CG, given the labeled lengths. 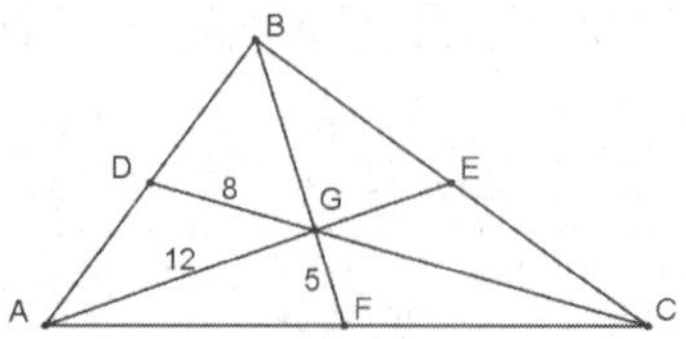	

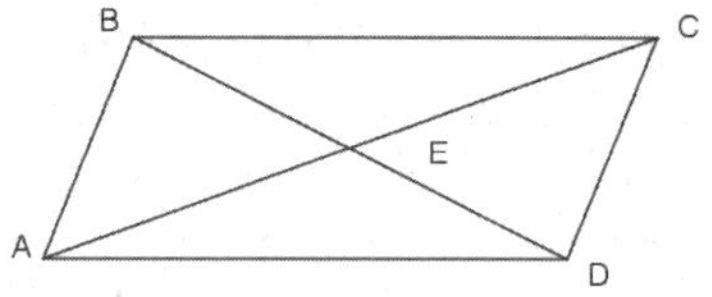

Lesson 34: Review of Assumptions

Problem Set

Use any of the assumptions, facts, and/or properties presented in the tables above to find x and/or y in each figure below. Justify your solutions.

1. Find the perimeter of parallelogram $ABCD$. Justify your solution.

2. $AC = 34$
   $AB = 26$
   $BD = 28$

   Given parallelogram $ABCD$, find the perimeter of $\triangle CED$. Justify your solution.

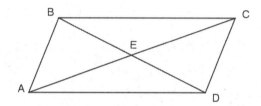

3. $XY = 12$
   $XZ = 20$
   $ZY = 24$

   F, G, and H are midpoints of the sides on which they are located. Find the perimeter of $\triangle FGH$. Justify your solution.

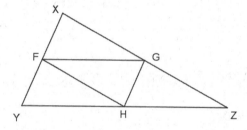

4. $ABCD$ is a parallelogram with $AE = CF$.
   Prove that $DEBF$ is a parallelogram.

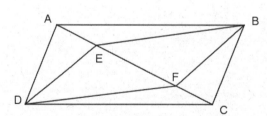

5. C is the centroid of $\triangle RST$.
   $RC = 16, CL = 10, TJ = 21$

   $SC = $ _____
   $TC = $ _____
   $KC = $ _____

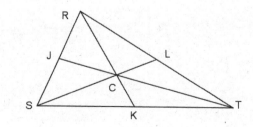

This page intentionally left blank

Student Edition

Eureka Math
Geometry
Module 2

Special thanks go to the Gordan A. Cain Center and to the Department of Mathematics at Louisiana State University for their support in the development of *Eureka Math*.

Published by Great Minds

Copyright © 2015 Great Minds. All rights reserved. No part of this work may be reproduced or used in any form or by any means — graphic, electronic, or mechanical, including photocopying or information storage and retrieval systems — without written permission from the copyright holder. "Great Minds" and "Eureka Math" are registered trademarks of Great Minds.

Printed in the U.S.A.
This book may be purchased from the publisher at eureka-math.org
10 9 8 7 6 5 4 3
ISBN 978-1-63255-327-0

Lesson 1: Scale Drawings

Classwork

Opening Exercise

Above is a picture of a bicycle. Which of the images below appears to be a well-scaled image of the original? Why?

A STORY OF FUNCTIONS   Lesson 1   M2

GEOMETRY

Example 1

Use construction tools to create a scale drawing of $\triangle ABC$ with a scale factor of $r = 2$.

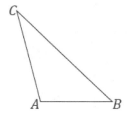

Exercise 1

1. Use construction tools to create a scale drawing of $\triangle DEF$ with a scale factor of $r = 3$. What properties does your scale drawing share with the original figure? Explain how you know.

Lesson 1: Scale Drawings

Example 2

Use construction tools to create a scale drawing of $\triangle XYZ$ with a scale factor of $r = \frac{1}{2}$.

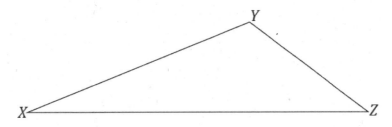

Exercises 2–4

2. Use construction tools to create a scale drawing of $\triangle PQR$ with a scale factor of $r = \frac{1}{4}$. What properties do the scale drawing and the original figure share? Explain how you know.

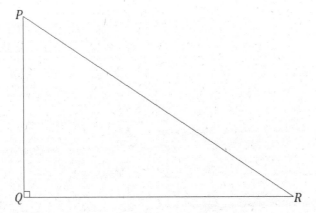

3. Triangle EFG is provided below, and one angle of scale drawing $\triangle E'F'G'$ is also provided. Use construction tools to complete the scale drawing so that the scale factor is $r = 3$. What properties do the scale drawing and the original figure share? Explain how you know.

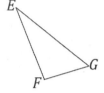

4. Triangle ABC is provided below, and one side of scale drawing $\triangle A'B'C'$ is also provided. Use construction tools to complete the scale drawing and determine the scale factor.

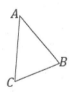

Lesson 1: Scale Drawings

Problem Set

1. Use construction tools to create a scale drawing of △ABC with a scale factor of $r = 3$.

2. Use construction tools to create a scale drawing of △ABC with a scale factor of $r = \frac{1}{2}$.

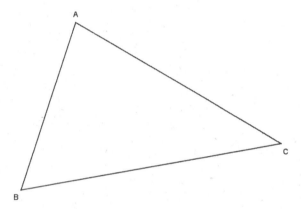

3. Triangle EFG is provided below, and one angle of scale drawing △$E'F'G'$ is also provided. Use construction tools to complete a scale drawing so that the scale factor is $r = 2$.

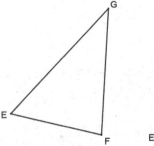

Lesson 1: Scale Drawings

4. Triangle MTC is provided below, and one angle of scale drawing $\triangle M'T'C'$ is also provided. Use construction tools to complete a scale drawing so that the scale factor is $r = \frac{1}{4}$.

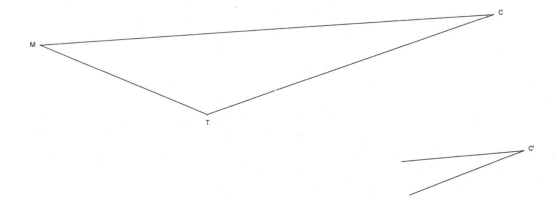

5. Triangle ABC is provided below, and one side of scale drawing $\triangle A'B'C'$ is also provided. Use construction tools to complete the scale drawing and determine the scale factor.

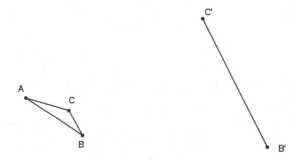

6. Triangle XYZ is provided below, and one side of scale drawing $\triangle X'Y'Z'$ is also provided. Use construction tools to complete the scale drawing and determine the scale factor.

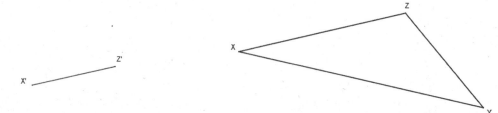

7. Quadrilateral $GHIJ$ is a scale drawing of quadrilateral $ABCD$ with scale factor r. Describe each of the following statements as always true, sometimes true, or never true, and justify your answer.

   a. $AB = GH$
   b. $m\angle ABC = m\angle GHI$
   c. $\dfrac{AB}{GH} = \dfrac{BC}{HI}$
   d. Perimeter$(GHIJ) = r \cdot$ Perimeter$(ABCD)$
   e. Area$(GHIJ) = r \cdot$ Area$(ABCD)$ where $r \neq 1$
   f. $r < 0$

This page intentionally left blank

Lesson 2: Making Scale Drawings Using the Ratio Method

Classwork

Opening Exercise

Based on what you recall from Grade 8, describe what a *dilation* is.

Example 1

Create a scale drawing of the figure below using the ratio method about center O and scale factor $r = \frac{1}{2}$.

Step 1. Draw a ray beginning at O through each vertex of the figure.

Step 2. Dilate each vertex along the appropriate ray by scale factor $r = \frac{1}{2}$. Use the ruler to find the midpoint between O and D and then each of the other vertices. Label each respective midpoint with prime notation (e.g., D').

Step 3. Join vertices in the way they are joined in the original figure (e.g., segment $A'B'$ corresponds to segment AB).

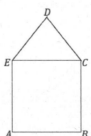

Exercise 1

1. Create a scale drawing of the figure below using the ratio method about center O and scale factor $r = \frac{3}{4}$. Verify that the resulting figure is in fact a scale drawing by showing that corresponding side lengths are in constant proportion and the corresponding angles are equal in measurement.

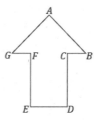

Example 2

a. Create a scale drawing of the figure below using the ratio method about center O and scale factor $r = 3$.

Step 1. Draw a ray beginning at O through each vertex of the figure.

Step 2. Use your ruler to determine the location of A' on $\overrightarrow{OA}$; A' should be three times as far from O as A. Determine the locations of B' and C' in the same way along the respective rays.

Step 3. Draw the corresponding line segments (e.g., segment $A'B'$ corresponds to segment AB).

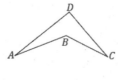

Lesson 2: Making Scale Drawings Using the Ratio Method

b. Locate a point X so that it lies between endpoints A and B on segment AB of the original figure in part (a). Use the ratio method to locate X' on the scale drawing in part (a).

c. Imagine a dilation of the same figure as in parts (a) and (b). What if the ray from the center passed through two distinct points, such as B and D? What does that imply about the locations of B' and D'?

Exercises 2–6

2. $\triangle A'B'C'$ is a scale drawing of $\triangle ABC$ drawn by using the ratio method. Use your ruler to determine the location of the center O used for the scale drawing.

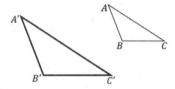

A STORY OF FUNCTIONS Lesson 2 M2

GEOMETRY

3. Use the figure below with center O and a scale factor of $r = \frac{5}{2}$ to create a scale drawing. Verify that the resulting figure is in fact a scale drawing by showing that corresponding side lengths are in constant proportion and that the corresponding angles are equal in measurement.

4. Summarize the steps to create a scale drawing by the ratio method. Be sure to describe all necessary parameters to use the ratio method.

5. A clothing company wants to print the face of the Statue of Liberty on a T-shirt. The length of the face from the top of the forehead to the chin is 17 feet, and the width of the face is 10 feet. Given that a medium-sized T-shirt has a length of 29 inches and a width of 20 inches, what dimensions of the face are needed to produce a scaled version that will fit on the T-shirt?

    a. What shape would you use to model the face of the statue?

b. Knowing that the maximum width of the T-shirt is 20 inches, what scale factor is needed to make the width of the face fit on the shirt?

c. What scale factor should be used to scale the length of the face? Explain.

d. Using the scale factor identified in part (c), what is the scaled length of the face? Will it fit on the shirt?

e. Identify the scale factor you would use to ensure that the face of the statue was in proportion and would fit on the T-shirt. Identify the dimensions of the face that will be printed on the shirt.

f. The T-shirt company wants the width of the face to be no smaller than 10 inches. What scale factors could be used to create a scaled version of the face that meets this requirement?

g. If it costs the company $0.005 for each square inch of print on a shirt, what are the maximum and minimum costs for printing the face of the Statue of Liberty on one T-shirt?

6. Create your own scale drawing using the ratio method. In the space below:
   a. Draw an original figure.
   b. Locate and label a center of dilation O.
   c. Choose a scale factor r.
   d. Describe your dilation using appropriate notation.
   e. Complete a scale drawing using the ratio method.

   Show all measurements and calculations to confirm that the new figure is a scale drawing. The work here will be your answer key.

   Next, trace your original figure onto a fresh piece of paper. Trade the traced figure with a partner. Provide your partner with the dilation information. Each partner should complete the other's scale drawing. When finished, check all work for accuracy against your answer key.

Problem Set

1. Use the ratio method to create a scale drawing about center O with a scale factor of $r = \frac{1}{4}$. Use a ruler and protractor to verify that the resulting figure is in fact a scale drawing by showing that corresponding side lengths are in constant proportion and the corresponding angles are equal in measurement.

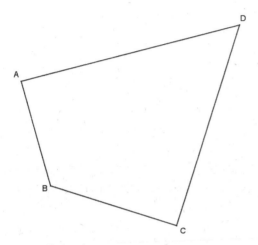

2. Use the ratio method to create a scale drawing about center O with a scale factor of $r = 2$. Verify that the resulting figure is in fact a scale drawing by showing that corresponding side lengths are in constant proportion and that the corresponding angles are equal in measurement.

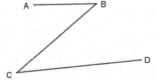

3. Use the ratio method to create two scale drawings: $D_{O,2}$ and $D_{P,2}$. Label the scale drawing with respect to center O as $\triangle A'B'C'$ and the scale drawing with respect to center P as $\triangle A''B''C''$.

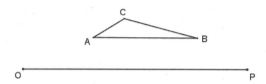

What do you notice about the two scale drawings?

What rigid motion can be used to map $\triangle A'B'C'$ onto $\triangle A''B''C''$?

4. Sara found a drawing of a triangle that appears to be a scale drawing. Much of the drawing has faded, but she can see the drawing and construction lines in the diagram below. If we assume the ratio method was used to construct $\triangle A'B'C'$ as a scale model of $\triangle ABC$, can you find the center O, the scale factor r, and locate $\triangle ABC$?

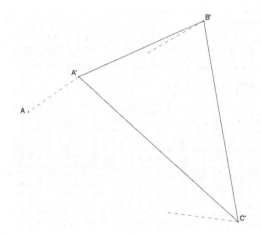

Lesson 2: Making Scale Drawings Using the Ratio Method

5. Quadrilateral $A'''B'''C'''D'''$ is one of a sequence of three scale drawings of quadrilateral $ABCD$ that were all constructed using the ratio method from center O. Find the center O, each scale drawing in the sequence, and the scale factor for each scale drawing. The other scale drawings are quadrilaterals $A'B'C'D'$ and $A''B''C''D''$.

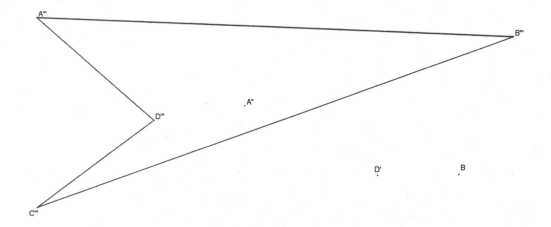

6. Maggie has a rectangle drawn in the corner of an $8\frac{1}{2}$-inch by 11-inch sheet of printer paper as shown in the diagram. To cut out the rectangle, Maggie must make two cuts. She wants to scale the rectangle so that she can cut it out using only one cut with a paper cutter.

   a. What are the dimensions of Maggie's scaled rectangle, and what is its scale factor from the original rectangle?

   b. After making the cut for the scaled rectangle, is there enough material left to cut another identical rectangle? If so, what is the area of scrap per sheet of paper?

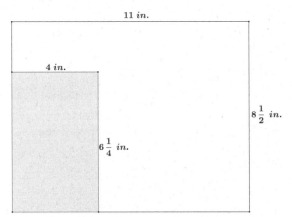

Lesson 2: Making Scale Drawings Using the Ratio Method

This page intentionally left blank

Lesson 3: Making Scale Drawings Using the Parallel Method

Classwork

Opening Exercise

Dani dilated △ ABC from center O, resulting in △ $A'B'C'$. She says that she completed the drawing using parallel lines. How could she have done this? Explain.

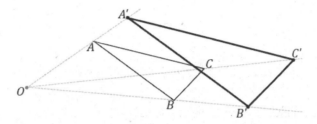

Example 1

a. Use a ruler and setsquare to draw a line through C parallel to $\overline{AB}$. What ensures that the line drawn is parallel to $\overline{AB}$?

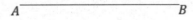

b. Use a ruler and setsquare to draw a parallelogram $ABCD$ around $\overline{AB}$ and point C.

Example 2

Use the figure below with center O and a scale factor of $r = 2$ and the following steps to create a scale drawing using the parallel method.

Step 1. Draw a ray beginning at O through each vertex of the figure.

Step 2. Select one vertex of the scale drawing to locate; we have selected A'. Locate A' on $\overrightarrow{OA}$ so that $OA' = 2OA$.

Step 3. Align the setsquare and ruler as in the image below; one leg of the setsquare should line up with side AB, and the perpendicular leg should be flush against the ruler.

Step 4. Slide the setsquare along the ruler until the edge of the setsquare passes through A'. Then, along the perpendicular leg of the setsquare, draw the segment through A' that is parallel to $\overline{AB}$ until it intersects with $\overrightarrow{OB}$, and label this point B'.

Step 5. Continue to create parallel segments to determine each successive vertex point. In this particular case, the setsquare has been aligned with $\overline{AC}$. This is done because, in trying to create a parallel segment from $\overline{BC}$, the parallel segment was not reaching B'. This could be remedied with a larger setsquare and longer ruler, but it is easily avoided by working on the segment parallel to $\overline{AC}$ instead.

Step 6. Use your ruler to join the final two unconnected vertices.

A STORY OF FUNCTIONS — Lesson 3 M2

GEOMETRY

Exercises

1. With a ruler and setsquare, use the parallel method to create a scale drawing of $WXYZ$ by the parallel method. W' has already been located for you. Determine the scale factor of the scale drawing. Verify that the resulting figure is in fact a scale drawing by showing that corresponding side lengths are in constant proportion and that corresponding angles are equal in measurement.

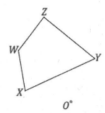

2. With a ruler and setsquare, use the parallel method to create a scale drawing of $DEFG$ about center O with scale factor $r = \frac{1}{2}$. Verify that the resulting figure is in fact a scale drawing by showing that corresponding side lengths are in constant proportion and that the corresponding angles are equal in measurement.

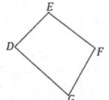

Lesson 3: Making Scale Drawings Using the Parallel Method

3. With a ruler and setsquare, use the parallel method to create a scale drawing of pentagon $PQRST$ about center O with scale factor $\frac{5}{2}$. Verify that the resulting figure is in fact a scale drawing by showing that corresponding side lengths are in constant proportion and that corresponding angles are equal in measurement.

A STORY OF FUNCTIONS   Lesson 3 M2
                                                                                          GEOMETRY

Problem Set

1. With a ruler and setsquare, use the parallel method to create a scale drawing of the figure about center O. One vertex of the scale drawing has been provided for you.

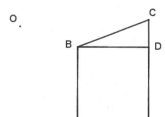

Determine the scale factor. Verify that the resulting figure is in fact a scale drawing by showing that corresponding side lengths are in constant proportion and that the corresponding angles are equal in measurement.

2. With a ruler and setsquare, use the parallel method to create a scale drawing of the figure about center O and scale factor $r = \frac{1}{3}$. Verify that the resulting figure is in fact a scale drawing by showing that corresponding side lengths are in constant proportion and the corresponding angles are equal in measurement.

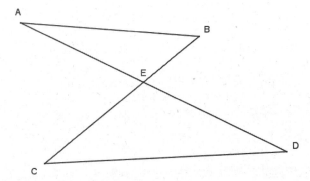

Lesson 3: Making Scale Drawings Using the Parallel Method

3. With a ruler and setsquare, use the parallel method to create the following scale drawings about center O: (1) first use a scale a factor of 2 to create $\triangle A'B'C'$, (2) then, with respect to $\triangle A'B'C'$, use a scale factor of $\frac{2}{3}$ to create scale drawing $\triangle A''B''C''$. Calculate the scale factor for $\triangle A''B''C''$ as a scale drawing of $\triangle ABC$. Use angle and side length measurements and the appropriate proportions to verify your answer.

O ·

4. Follow the direction in each part below to create three scale drawings of $\triangle ABC$ using the parallel method.

    a. With the center at vertex A, make a scale drawing of $\triangle ABC$ with a scale factor of $\frac{3}{2}$.

    b. With the center at vertex B, make a scale drawing of $\triangle ABC$ with a scale factor of $\frac{3}{2}$.

    c. With the center at vertex C, make a scale drawing of $\triangle ABC$ with a scale factor of $\frac{3}{2}$.

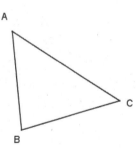

    d. What conclusions can be drawn about all three scale drawings from parts (a)–(c)?

5. Use the parallel method to make a scale drawing of the line segments in the following figure using the given W', the image of vertex W, from center O. Determine the scale factor.

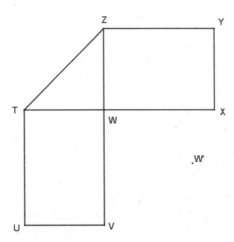

Use your diagram from Problem 1 to answer this question.

6. If we switch perspective and consider the original drawing $ABCDE$ to be a scale drawing of the constructed image $A'B'C'D'E'$, what would the scale factor be?

This page intentionally left blank

Lesson 4: Comparing the Ratio Method with the Parallel Method

Classwork

Today, our goal is to show that the parallel method and the ratio method are equivalent; that is, given a figure in the plane and a scale factor $r > 0$, the scale drawing produced by the parallel method is congruent to the scale drawing produced by the ratio method. We start with two easy exercises about the areas of two triangles whose bases lie on the same line, which helps show that the two methods are equivalent.

Opening Exercise

a. Suppose two triangles, $\triangle ABC$ and $\triangle ABD$, share the same base $\overline{AB}$ such that points C and D lie on a line parallel to $\overleftrightarrow{AB}$. Show that their areas are equal, that is, Area($\triangle ABC$) = Area($\triangle ABD$). (Hint: Why are the altitudes of each triangle equal in length?)

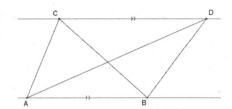

b. Suppose two triangles have different-length bases, $\overline{AB}$ and $\overline{AB'}$, that lie on the same line. Furthermore, suppose they both have the same vertex C opposite these bases. Show that the value of the ratio of their areas is equal to the value of the ratio of the lengths of their bases, that is,

$$\frac{\text{Area}(\triangle ABC)}{\text{Area}(\triangle AB'C)} = \frac{AB}{AB'}.$$

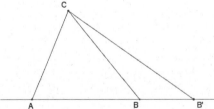

Discussion

To show that the parallel and ratio methods are equivalent, we need only look at one of the simplest versions of a scale drawing: scaling segments. First, we need to show that the scale drawing of a segment generated by the parallel method is the same segment that the ratio method would have generated and vice versa. That is,

The parallel method $\Rightarrow$ The ratio method,

and

The ratio method $\Rightarrow$ The parallel method.

The first implication above can be stated as the following theorem:

PARALLEL $\Rightarrow$ RATIO THEOREM: Given $\overline{AB}$ and point O not on $\overleftrightarrow{AB}$, construct a scale drawing of $\overline{AB}$ with scale factor $r > 0$ using the parallel method: Let $A' = D_{O,r}(A)$, and ℓ be the line parallel to $\overleftrightarrow{AB}$ that passes through A'. Let B' be the point where $\overrightarrow{OB}$ intersects ℓ. Then B' is the same point found by the ratio method, that is, $B' = D_{O,r}(B)$.

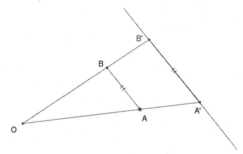

PROOF: We prove the case when $r > 1$; the case when $0 < r < 1$ is the same but with a different picture. Construct two line segments $\overline{BA'}$ and $\overline{AB'}$ to form two triangles $\triangle BAB'$ and $\triangle BAA'$, labeled as T_1 and T_2, respectively, in the picture below.

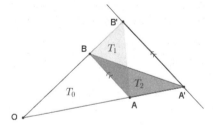

The areas of these two triangles are equal,

$$\text{Area}(T_1) = \text{Area}(T_2),$$

by Exercise 1. Why? Label $\triangle OAB$ by T_0. Then $\text{Area}(\triangle OA'B) = \text{Area}(\triangle OB'A)$ because areas add:

$$\text{Area}(\triangle OA'B) = \text{Area}(T_0) + \text{Area}(T_2)$$
$$= \text{Area}(T_0) + \text{Area}(T_1)$$
$$= \text{Area}(\triangle OB'A).$$

Next, we apply Exercise 2 to two sets of triangles: (1) T_0 and $\triangle OA'B$ and (2) T_0 and $\triangle OB'A$.

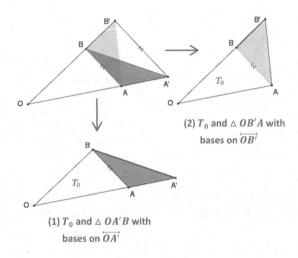

Therefore,

$$\frac{\text{Area}(\triangle OA'B)}{\text{Area}(T_0)} = \frac{OA'}{OA}, \text{ and}$$

$$\frac{\text{Area}(\triangle OB'A)}{\text{Area}(T_0)} = \frac{OB'}{OB}.$$

Since $\text{Area}(\triangle OA'B) = \text{Area}(\triangle OB'A)$, we can equate the fractions: $\frac{OA'}{OA} = \frac{OB'}{OB}$. Since r is the scale factor used in dilating $\overline{OA}$ to $\overline{OA'}$, we know that $\frac{OA'}{OA} = r$; therefore, $\frac{OB'}{OB} = r$, or $OB' = r \cdot OB$. This last equality implies that B' is the dilation of B from O by scale factor r, which is what we wanted to prove.

Next, we prove the reverse implication to show that both methods are equivalent to each other.

RATIO ⟹ PARALLEL THEOREM: Given $\overline{AB}$ and point O not on $\overleftrightarrow{AB}$, construct a scale drawing $\overline{A'B'}$ of $\overline{AB}$ with scale factor $r > 0$ using the ratio method (i.e., find $A' = D_{O,r}(A)$ and $B' = D_{O,r}(B)$, and draw $\overline{A'B'}$). Then B' is the same as the point found using the parallel method.

PROOF: Since both the ratio method and the parallel method start with the same first step of setting $A' = D_{O,r}(A)$, the only difference between the two methods is in how the second point is found. If we use the parallel method, we construct the line ℓ parallel to $\overleftrightarrow{AB}$ that passes through A' and label the point where ℓ intersects $\overrightarrow{OB}$ by C. Then B' is the same as the point found using the parallel method if we can show that $C = B'$.

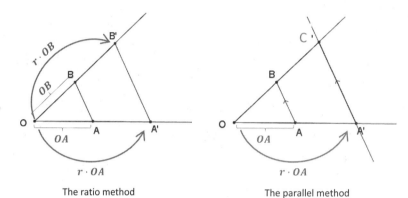

The ratio method   The parallel method

By the parallel ⟹ ratio theorem, we know that $C = D_{O,r}(B)$, that is, that C is the point on $\overrightarrow{OB}$ such that $OC = r \cdot OB$. But B' is also the point on $\overrightarrow{OB}$ such that $OB' = r \cdot OB$. Hence, they must be the same point.

The fact that the ratio and parallel methods are equivalent is often stated as the triangle side splitter theorem. To understand the triangle side splitter theorem, we need a definition:

SIDE SPLITTER: A line segment CD is said to split *the sides of* $\triangle\ OAB$ *proportionally* if C is a point on $\overline{OA}$, D is a point on $\overline{OB}$, and $\dfrac{OA}{OC} = \dfrac{OB}{OD}$ (or equivalently, $\dfrac{OC}{OA} = \dfrac{OD}{OB}$). We call line segment CD a *side* splitter.

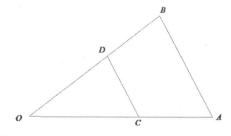

TRIANGLE SIDE SPLITTER THEOREM: A line segment splits two sides of a triangle proportionally if and only if it is parallel to the third side.

Restatement of the triangle *side splitter* theorem:

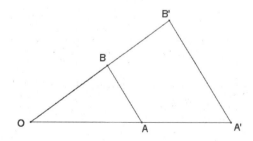

Lesson Summary

THE TRIANGLE SIDE SPLITTER THEOREM: A line segment splits two sides of a triangle proportionally if and only if it is parallel to the third side.

Problem Set

1. Use the diagram to answer each part below.

   a. Measure the segments in the figure below to verify that the proportion is true.
   $$\frac{OA'}{OA} = \frac{OB'}{OB}$$

   b. Is the proportion $\frac{OA}{OA'} = \frac{OB}{OB'}$ also true? Explain algebraically.

   c. Is the proportion $\frac{AA'}{OA'} = \frac{BB'}{OB'}$ also true? Explain algebraically.

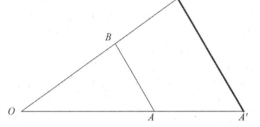

2. Given the diagram below, $AB = 30$, line ℓ is parallel to $\overline{AB}$, and the distance from $\overline{AB}$ to ℓ is 25. Locate point C on line ℓ such that $\triangle ABC$ has the greatest area. Defend your answer.

3. Given $\triangle XYZ$, $\overline{XY}$ and $\overline{YZ}$ are partitioned into equal-length segments by the endpoints of the dashed segments as shown. What can be concluded about the diagram?

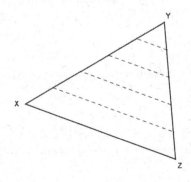

4. Given the diagram, $AC = 12$, $AB = 6$, $BE = 4$, $m\angle ACB = x°$, and $m\angle D = x°$, find CD.

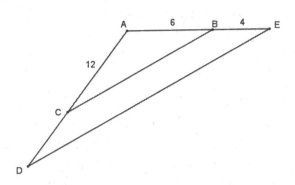

5. What conclusions can be drawn from the diagram shown to the right? Explain.

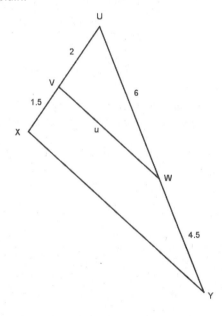

6. Parallelogram $PQRS$ is shown. Two triangles are formed by a diagonal within the parallelogram. Identify those triangles, and explain why they are guaranteed to have the same areas.

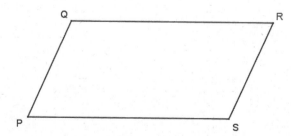

7. In the diagram to the right, $HI = 36$ and $GJ = 42$. If the ratio of the areas of the triangles is $\dfrac{\text{Area } \triangle GHI}{\text{Area } \triangle JHI} = \dfrac{5}{9}$, find JH, GH, GI, and JI.

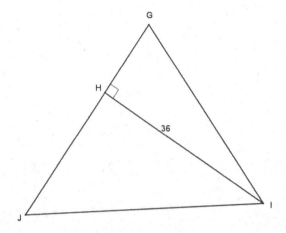

This page intentionally left blank

Lesson 5: Scale Factors

Classwork

Opening Exercise

Quick Write: Describe how a figure is transformed under a dilation with a scale factor $= 1$, $r > 1$, and $0 < r < 1$.

Discussion

DILATION THEOREM: If a dilation with center O and scale factor r sends point P to P' and Q to Q', then $|P'Q'| = r|PQ|$. Furthermore, if $r \neq 1$ and $O, P,$ and Q are the vertices of a triangle, then $\overleftrightarrow{PQ} \| \overleftrightarrow{P'Q'}$.

Now consider the dilation theorem when $O, P,$ and Q are the vertices of $\triangle OPQ$. Since P' and Q' come from a dilation with scale factor r and center O, we have $\dfrac{OP'}{OP} = \dfrac{OQ'}{OQ} = r$.

There are two cases that arise; recall what you wrote in your Quick Write. We must consider the case when $r > 1$ and when $0 < r < 1$. Let's begin with the latter.

Dilation Theorem Proof, Case 1

Statements	Reasons/Explanations
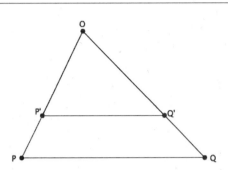 1. A dilation with center O and scale factor r sends point P to P' and Q to Q'.	1.
2. $\dfrac{OP'}{OP} = \dfrac{OQ'}{OQ} = r$	2.
3. $\overleftrightarrow{PQ} \parallel \overleftrightarrow{P'Q'}$	3.
4. A dilation with center P and scale factor $\dfrac{PP'}{PO}$ sends point O to P' and point Q to R. Draw $\overline{P'R}$. 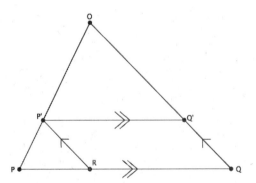	4.
5. $\overleftrightarrow{P'R} \parallel \overleftrightarrow{OQ'}$	5.

Lesson 5: Scale Factors

6. $RP'Q'Q$ is a parallelogram.

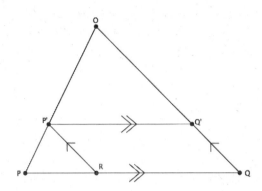

7. $RQ = P'Q'$

8. $\dfrac{RQ}{PQ} = \dfrac{P'O}{PO}$

9. $\dfrac{RQ}{PQ} = r$

10. $RQ = r \cdot PQ$

11. $P'Q' = r \cdot PQ$

6.

7.

8.

9.

10.

11.

Exercises

1. Prove Case 2: If O, P, and Q are the vertices of a triangle and $r > 1$, show that (a) $\overleftrightarrow{PQ} \parallel \overleftrightarrow{P'Q'}$ and (b) $P'Q' = rPQ$. Use the diagram below when writing your proof.

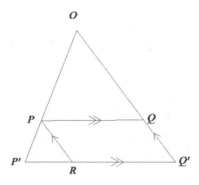

2.
   a. Produce a scale drawing of $\triangle LMN$ using either the ratio or parallel method with point M as the center and a scale factor of $\frac{3}{2}$.

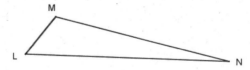

b. Use the dilation theorem to predict the length of $\overline{L'N'}$, and then measure its length directly using a ruler.

c. Does the dilation theorem appear to hold true?

3. Produce a scale drawing of △ XYZ with point X as the center and a scale factor of $\frac{1}{4}$. Use the dilation theorem to predict $Y'Z'$, and then measure its length directly using a ruler. Does the dilation theorem appear to hold true?

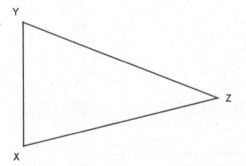

4. Given the diagram below, determine if △ DEF is a scale drawing of △ DGH. Explain why or why not.

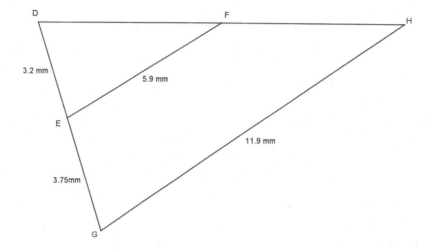

Problem Set

1. △AB'C' is a dilation of △ABC from vertex A, and CC' = 2. Use the given information in each part and the diagram to find B'C'.

   a. $AB = 9$, $AC = 4$, and $BC = 7$
   b. $AB = 4$, $AC = 9$, and $BC = 7$
   c. $AB = 7$, $AC = 9$, and $BC = 4$
   d. $AB = 7$, $AC = 4$, and $BC = 9$
   e. $AB = 4$, $AC = 7$, and $BC = 9$
   f. $AB = 9$, $AC = 7$, and $BC = 4$

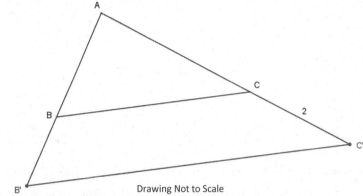

Drawing Not to Scale

2. Given the diagram, $\angle CAB \cong \angle CFE$. Find AB.

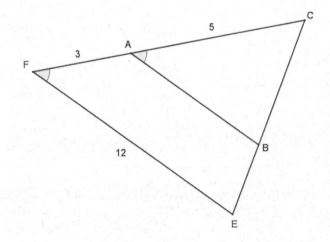

3. Use the diagram to answer each part below.

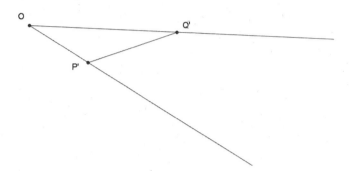

   a. $\triangle OP'Q'$ is the dilated image of $\triangle OPQ$ from point O with a scale factor of $r > 1$. Draw a possible $\overline{PQ}$.
   b. $\triangle OP''Q''$ is the dilated image of $\triangle OPQ$ from point O with a scale factor $k > r$. Draw a possible $\overline{P''Q''}$.

4. Given the diagram to the right, $\overline{RS} \parallel \overline{PQ}$, Area ($\triangle RST$) = 15 units², and Area($\triangle OSR$) = 21 units², find RS.

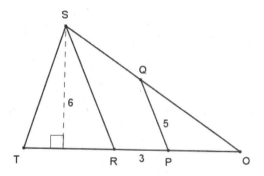

5. Using the information given in the diagram and $UX = 9$, find Z on $\overline{XU}$ such that $\overline{YZ}$ is an altitude. Then, find YZ and XZ.

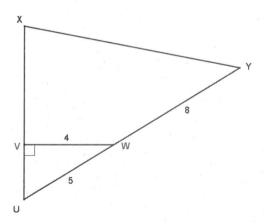

Lesson 6: Dilations as Transformations of the Plane

Classwork

Exercises 1–6

1. Find the center and the angle of the rotation that takes $\overline{AB}$ to $\overline{A'B'}$. Find the image P' of point P under this rotation.

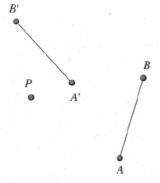

2. In the diagram below, △ $B'C'D'$ is the image of △ BCD after a rotation about a point A. What are the coordinates of A, and what is the degree measure of the rotation?

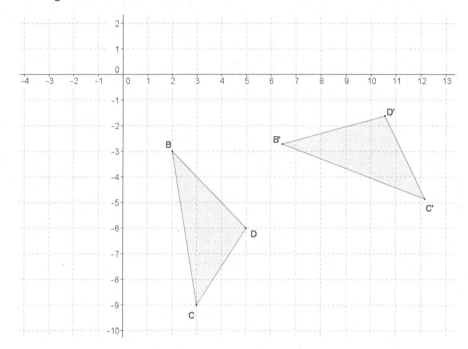

3. Find the line of reflection for the reflection that takes point A to point A'. Find the image P' under this reflection.

4. Quinn tells you that the vertices of the image of quadrilateral $CDEF$ reflected over the line representing the equation $y = -\frac{3}{2}x + 2$ are the following: $C'(-2,3)$, $D'(0,0)$, $E'(-3,-3)$, and $F'(-3,3)$. Do you agree or disagree with Quinn? Explain.

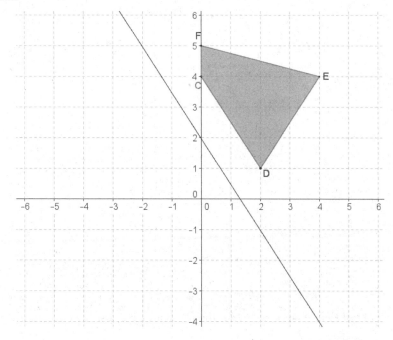

5. A translation takes A to A'. Find the image P' and pre-image P'' of point P under this translation. Find a vector that describes the translation.

6. The point C' is the image of point C under a translation of the plane along a vector.

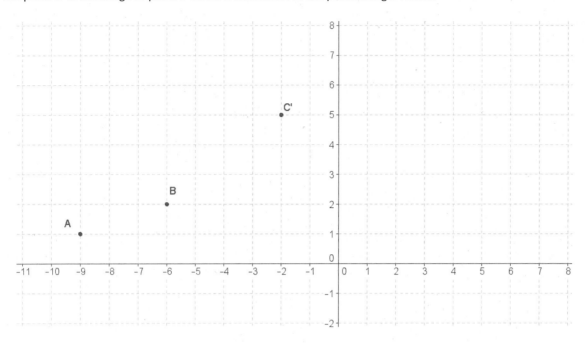

a. Find the coordinates of C if the vector used for the translation is $\overrightarrow{BA}$.

b. Find the coordinates of C if the vector used for the translation is $\overrightarrow{AB}$.

Exercises 7–9

7. A dilation with center O and scale factor r takes A to A' and B to B'. Find the center O, and estimate the scale factor r.

8. Given a center O, scale factor r, and points A and B, find the points $A' = D_{O,r}(A)$ and $B' = D_{O,r}(B)$. Compare length AB with length $A'B'$ by division; in other words, find $\dfrac{A'B'}{AB}$. How does this number compare to r?

$r = 3$

Lesson 6: Dilations as Transformations of the Plane

9. Given a center O, scale factor r, and points A, B, and C, find the points $A' = D_{O,r}(A)$, $B' = D_{O,r}(B)$, and $C' = D_{O,r}(C)$. Compare $m\angle ABC$ with $m\angle A'B'C'$. What do you find?

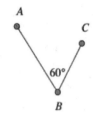

Lesson Summary

- There are two major classes of transformations: those that are distance preserving (translations, reflections, rotations) and those that are not (dilations).
- Like rigid motions, dilations involve a rule assignment for each point in the plane and also have inverse functions that return each dilated point back to itself.

Problem Set

1. In the diagram below, A' is the image of A under a single transformation of the plane. Use the given diagram to show your solutions to parts (a)–(d).

   a. Describe the translation that maps $A \to A'$, and then use the translation to locate P', the image of P.
   b. Describe the reflection that maps $A \to A'$, and then use the reflection to locate P', the image of P.
   c. Describe a rotation that maps $A \to A'$, and then use your rotation to locate P', the image of P.
   d. Describe a dilation that maps $A \to A'$, and then use your dilation to locate P', the image of P.

Lesson 6: Dilations as Transformations of the Plane

2. On the diagram below, O is a center of dilation, and $\overleftrightarrow{AD}$ is a line not through O. Choose two points B and C on $\overleftrightarrow{AD}$ between A and D.

   a. Dilate $A, B, C,$ and D from O using scale factor $r = \frac{1}{2}$. Label the images $A', B', C',$ and D', respectively.
   b. Dilate $A, B, C,$ and D from O using scale factor $r = 2$. Label the images $A'', B'', C'',$ and D'', respectively.
   c. Dilate $A, B, C,$ and D from O using scale factor $r = 3$. Label the images $A''', B''', C''',$ and D''', respectively.
   d. Draw a conclusion about the effect of a dilation on a line segment based on the diagram that you drew. Explain.

3. Write the inverse transformation for each of the following so that the composition of the transformation with its inverse maps a point to itself on the plane.
   a. $T_{\overrightarrow{AB}}$
   b. $r_{\overleftrightarrow{AB}}$
   c. $R_{C,45}$
   d. $D_{O,r}$

4. Given $U(1,3), V(-4,-4),$ and $W(-3,6)$ on the coordinate plane, perform a dilation of $\triangle UVW$ from center $O(0,0)$ with a scale factor of $\frac{3}{2}$. Determine the coordinates of images of points $U, V,$ and W, and describe how the coordinates of the image points are related to the coordinates of the pre-image points.

5. Points $B, C, D, E, F,$ and G are dilated images of A from center O with scale factors 2, 3, 4, 5, 6, and 7, respectively. Are points $Y, X, W, V, U, T,$ and S all dilated images of Z under the same respective scale factors? Explain why or why not.

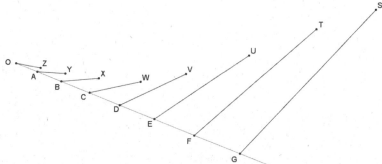

6. Find the center and scale factor that takes A to A' and B to B', if a dilation exists.

7. Find the center and scale factor that takes A to A' and B to B', if a dilation exists.

This page intentionally left blank

Lesson 7: How Do Dilations Map Segments?

Classwork

Opening Exercise

a. Is a dilated segment still a segment? If the segment is transformed under a dilation, explain how.

b. Dilate the segment PQ by a scale factor of 2 from center O.

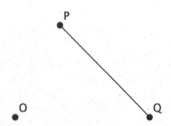

i. Is the dilated segment $P'Q'$ a segment?

ii. How, if at all, has the segment PQ been transformed?

Example 1

Case 1. Consider the case where the scale factor of dilation is $r = 1$. Does a dilation from center O map segment PQ to a segment $P'Q'$? Explain.

Example 2

Case 2. Consider the case where a line PQ contains the center of the dilation. Does a dilation from the center with scale factor $r \neq 1$ map the segment PQ to a segment $P'Q'$? Explain.

Example 3

Case 3. Consider the case where $\overleftrightarrow{PQ}$ does not contain the center O of the dilation, and the scale factor r of the dilation is not equal to 1; then, we have the situation where the key points O, P, and Q form $\triangle OPQ$. The scale factor not being equal to 1 means that we must consider scale factors such that $0 < r < 1$ and $r > 1$. However, the proofs for each are similar, so we focus on the case when $0 < r < 1$.

When we dilate points P and Q from center O by scale factor $0 < r < 1$, as shown, what do we know about points P' and Q'?

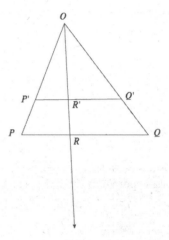

We use the fact that the line segment $P'Q'$ splits the sides of $\triangle OPQ$ proportionally and that the lines containing $\overline{PQ}$ and $\overline{P'Q'}$ are parallel to prove that a dilation maps segments to segments. Because we already know what happens when points P and Q are dilated, consider another point R that is on the segment PQ. After dilating R from center O by scale factor r to get the point R', does R' lie on the segment $P'Q'$?

Putting together the preliminary dilation theorem for segments with the dilation theorem, we get

DILATION THEOREM FOR SEGMENTS: A dilation $D_{O,r}$ maps a line segment PQ to a line segment $P'Q'$ sending the endpoints to the endpoints so that $P'Q' = rPQ$. Whenever the center O does not lie in line PQ and $r \neq 1$, we conclude $\overleftrightarrow{PQ} \parallel \overleftrightarrow{P'Q'}$. Whenever the center O lies in $\overleftrightarrow{PQ}$ or if $r = 1$, we conclude $\overleftrightarrow{PQ} = \overleftrightarrow{P'Q'}$.

As an aside, observe that a dilation maps parallel line segments to parallel line segments. Further, a dilation maps a directed line segment to a directed line segment that points in the same direction.

Example 4

Now look at the converse of the dilation theorem for segments: If $\overline{PQ}$ and $\overline{RS}$ are line segments of different lengths in the plane, then there is a dilation that maps one to the other if and only if $\overleftrightarrow{PQ} = \overleftrightarrow{RS}$ or $\overleftrightarrow{PQ} \parallel \overleftrightarrow{RS}$.

Based on Examples 2 and 3, we already know that a dilation maps a segment PQ to another line segment, say $\overline{RS}$, so that $\overleftrightarrow{PQ} = \overleftrightarrow{RS}$ (Example 2) or $\overleftrightarrow{PQ} \parallel \overleftrightarrow{RS}$ (Example 3). If $\overleftrightarrow{PQ} \parallel \overleftrightarrow{RS}$, then, because $\overline{PQ}$ and $\overline{RS}$ are different lengths in the plane, they are bases of a trapezoid, as shown.

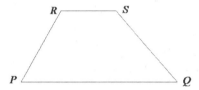

Since $\overline{PQ}$ and $\overline{RS}$ are segments of different lengths, then the non-base sides of the trapezoid are not parallel, and the lines containing them meet at a point O as shown.

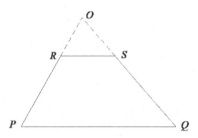

Recall that we want to show that a dilation maps $\overline{PQ}$ to $\overline{RS}$. Explain how to show it.

The case when the segments $\overline{PQ}$ and $\overline{RS}$ are such that $\overleftrightarrow{PQ} = \overleftrightarrow{RS}$ is left as an exercise.

Exercises 1–2

In the following exercises, you will consider the case where the segment and its dilated image belong to the same line, that is, when $\overline{PQ}$ and $\overline{RS}$ are such that $\overleftrightarrow{PQ} = \overleftrightarrow{RS}$.

1. Consider points P, Q, R, and S on a line, where $P = R$, as shown below. Show there is a dilation that maps $\overline{PQ}$ to $\overline{RS}$. Where is the center of the dilation?

2. Consider points P, Q, R, and S on a line as shown below where $PQ \neq RS$. Show there is a dilation that maps $\overline{PQ}$ to $\overline{RS}$. Where is the center of the dilation?

Lesson 7: How Do Dilations Map Segments?

Lesson Summary

- When a segment is dilated by a scale factor of $r = 1$, then the segment and its image would be the same length.
- When the points P and Q are on a line containing the center, then the dilated points P' and Q' are also collinear with the center producing an image of the segment that is a segment.
- When the points P and Q are not collinear with the center and the segment is dilated by a scale factor of $r \neq 1$, then the point P' lies on the ray OP' with $OP' = r \cdot OP$, and Q' lies on ray OQ with $OQ' = r \cdot OQ$.

Problem Set

1. Draw the dilation of parallelogram $ABCD$ from center O using the scale factor $r = 2$, and then answer the questions that follow.

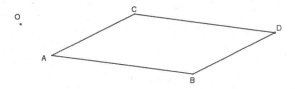

   a. Is the image $A'B'C'D'$ also a parallelogram? Explain.
   b. What do parallel lines seem to map to under a dilation?

2. Given parallelogram $ABCD$ with $A(-8,1)$, $B(2,-4)$, $C(-3,-6)$, and $D(-13,-1)$, perform a dilation of the plane centered at the origin using the following scale factors.

   a. Scale factor $\dfrac{1}{2}$
   b. Scale factor 2
   c. Are the images of parallel line segments under a dilation also parallel? Use your graphs to support your answer.

Lesson 7: How Do Dilations Map Segments?

3. In Lesson 7, Example 3, we proved that a line segment PQ, where O, P, and Q are the vertices of a triangle, maps to a line segment $P'Q'$ under a dilation with a scale factor $r < 1$. Using a similar proof, prove that for O not on $\overleftrightarrow{PQ}$, a dilation with center O and scale factor $r > 1$ maps a point R on $\overline{PQ}$ to a point R' on $\overleftrightarrow{PQ}$.

4. On the plane, $\overline{AB} \parallel \overline{A'B'}$ and $\overleftrightarrow{AB} \neq \overleftrightarrow{A'B'}$. Describe a dilation mapping $\overline{AB}$ to $\overline{A'B'}$. (Hint: There are 2 cases to consider.)

5. Only one of Figures A, B, or C below contains a dilation that maps A to A' and B to B'. Explain for each figure why the dilation does or does not exist. For each figure, assume that $\overleftrightarrow{AB} \neq \overleftrightarrow{A'B'}$.

   a.
   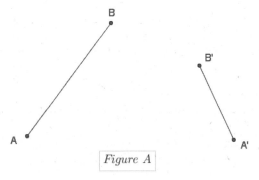

   b.
   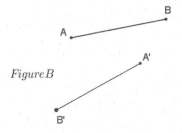

   c.
   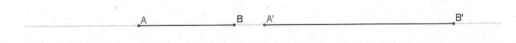

Lesson 7: How Do Dilations Map Segments?

This page intentionally left blank

Lesson 8: How Do Dilations Map Lines, Rays, and Circles?

Classwork

Opening Exercise

a. Is a dilated ray still a ray? If the ray is transformed under a dilation, explain how.

b. Dilate the $\overrightarrow{PQ}$ by a scale factor of 2 from center O.

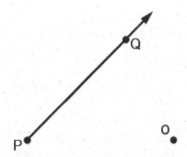

i. Is the figure $\overrightarrow{P'Q'}$ a ray?

ii. How, if at all, has the ray PQ been transformed?

iii. Will a ray always be mapped to a ray? Explain how you know.

Example 1

Will a dilation about center O and scale factor $r = 1$ map $\overrightarrow{PQ}$ to $\overrightarrow{P'Q'}$? Explain.

Example 2

The line that contains $\overrightarrow{PQ}$ does not contain point O. Does a dilation about center O and scale factor $r \neq 1$ map every point of $\overrightarrow{PQ}$ onto a point of $\overrightarrow{P'Q'}$?

Example 3

The line that contains $\overrightarrow{PQ}$ contains point O. Does a dilation about center O and scale factor r map $\overrightarrow{PQ}$ to $\overrightarrow{P'Q'}$?

a. Examine the case where the endpoint P of $\overrightarrow{PQ}$ coincides with the center O of the dilation.

b. Examine the case where the endpoint P of $\overrightarrow{PQ}$ is between O and Q on the line containing O, P, and Q.

c. Examine the remaining case where the center O of the dilation and point Q are on the same side of P on the line containing O, P, and Q.

Example 5

Does a dilation about a center O and scale factor r map a circle of radius R onto another circle?

a. Examine the case where the center of the dilation coincides with the center of the circle.

b. Examine the case where the center of the dilation is not the center of the circle; we call this the *general case*.

Lesson 8: How Do Dilations Map Lines, Rays, and Circles?

A STORY OF FUNCTIONS
Lesson 8 M2
GEOMETRY

Lesson Summary

- **DILATION THEOREM FOR RAYS**: A dilation maps a ray to a ray sending the endpoint to the endpoint.
- **DILATION THEOREM FOR LINES**: A dilation maps a line to a line. If the center O of the dilation lies on the line or if the scale factor r of the dilation is equal to 1, then the dilation maps the line to the same line. Otherwise, the dilation maps the line to a parallel line.
- **DILATION THEOREM FOR CIRCLES**: A dilation maps a circle to a circle and maps the center to the center.

Problem Set

1. In Lesson 8, Example 2, you proved that a dilation with a scale factor $r > 1$ maps a ray PQ to a ray $P'Q'$. Prove the remaining case that a dilation with scale factor $0 < r < 1$ maps a ray PQ to a ray $P'Q'$.

   Given the dilation $D_{O,r}$, with $0 < r < 1$ maps P to P' and Q to Q', prove that $D_{O,r}$ maps $\overrightarrow{PQ}$ to $\overrightarrow{P'Q'}$.

2. In the diagram below, $\overrightarrow{A'B'}$ is the image of $\overrightarrow{AB}$ under a dilation from point O with an unknown scale factor; A maps to A', and B maps to B'. Use direct measurement to determine the scale factor r, and then find the center of dilation O.

3. Draw a line $\overleftrightarrow{AB}$, and dilate points A and B from center O where O is not on $\overleftrightarrow{AB}$. Use your diagram to explain why a line maps to a line under a dilation with scale factor r.

4. Let $\overline{AB}$ be a line segment, and let m be a line that is the perpendicular bisector of $\overline{AB}$. If a dilation with scale factor r maps $\overline{AB}$ to $\overline{A'B'}$ (sending A to A' and B to B') and also maps line m to line m', show that line m' is the perpendicular bisector of $\overline{A'B'}$.

5. Dilate circle C with radius CA from center O with a scale factor $r = \frac{1}{2}$.

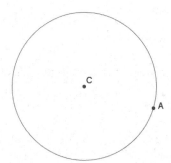

6. In the picture below, the larger circle is a dilation of the smaller circle. Find the center of dilation O.

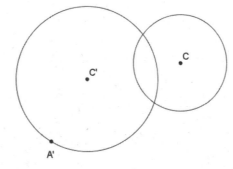

This page intentionally left blank

Lesson 9: How Do Dilations Map Angles?

Classwork

Exploratory Challenge/Exercises 1–4

1. How do dilations map triangles?

    a. Make a conjecture.

    b. Verify your conjecture by experimenting with diagrams and directly measuring angles and lengths of segments.

2. How do dilations map rectangles?

    a. Make a conjecture.

    b. Verify your conjecture by experimenting with diagrams and directly measuring angles and lengths of segments.

3. How do dilations map squares?

   a. Make a conjecture.

   b. Verify your conjecture by experimenting with diagrams and directly measuring angles and lengths of segments.

4. How do dilations map regular polygons?

   a. Make a conjecture.

   b. Verify your conjecture by experimenting with diagrams and directly measuring angles and lengths of segments.

Exercises 5–6

5. Recall what you learned about parallel lines cut by a transversal, specifically about the angles that are formed.

6. A dilation from center O by scale factor r maps $\angle BAC$ to $\angle B'A'C'$. Show that $m\angle BAC = m\angle B'A'C'$.

Discussion

The dilation theorem for angles is as follows:

DILATION THEOREM: A dilation from center O and scale factor r maps an angle to an angle of equal measure.

We have shown this when the angle and its image intersect at a single point, and that point of intersection is not the vertex of the angle.

A STORY OF FUNCTIONS
Lesson 9 M2
GEOMETRY

Lesson Summary

- Dilations map angles to angles of equal measure.
- Dilations map polygonal figures to polygonal figures whose angles are equal in measure to the corresponding angles of the original figure and whose side lengths are equal to the corresponding side lengths multiplied by the scale factor.

Problem Set

1. Shown below is $\triangle ABC$ and its image $\triangle A'B'C'$ after it has been dilated from center O by scale factor $r = \frac{5}{2}$. Prove that the dilation maps $\triangle ABC$ to $\triangle A'B'C'$ so that $m\angle A = m\angle A'$, $m\angle B = m\angle B'$, and $m\angle C = m\angle C'$.

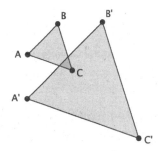

2. Explain the effect of a dilation with scale factor r on the length of the base and height of a triangle. How is the area of the dilated image related to the area of the pre-image?

3. Dilate trapezoid $ABDE$ from center O using a scale factor of $r = \frac{1}{2}$.

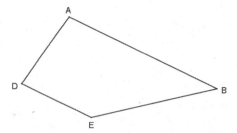

Lesson 9: How Do Dilations Map Angles?

4. Dilate kite $ABCD$ from center O using a scale factor $r = 1\frac{1}{2}$.

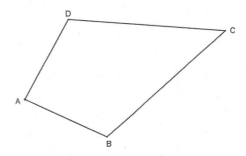

5. Dilate hexagon $DEFGHI$ from center O using a scale factor of $r = \frac{1}{4}$.

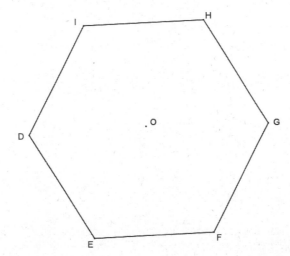

6. Examine the dilations that you constructed in Problems 2–5, and describe how each image compares to its pre-image under the given dilation. Pay particular attention to the sizes of corresponding angles and the lengths of corresponding sides.

This page intentionally left blank

Lesson 10: Dividing the King's Foot into 12 Equal Pieces

Classwork

Opening Exercise

Use a compass to mark off equally spaced points C, D, E, and F so that $\overline{AB}$, $\overline{BC}$, $\overline{CD}$, $\overline{DE}$, and $\overline{EF}$ are equal in length. Describe the steps you took.

Exploratory Challenge 1

Divide segment AB into three segments of equal lengths.

Exercise 1

Divide segment AB into five segments of equal lengths.

Exploratory Challenge 2

Divide segment AB into four segments of equal length.

Exercise 2

On a piece of poster paper, draw a segment AB with a measurement of 1 foot. Use the dilation method to divide $\overline{AB}$ into twelve equal-length segments, or into 12 inches.

Lesson Summary

SIDE SPLITTER METHOD: If $\overline{AB}$ is a line segment, construct a ray AA_1, and mark off n equally spaced points using a compass of fixed radius to get points $A = A_0, A_1, A_2, \cdots, A_n$. Construct $\overline{A_n B}$ that is a side of $\triangle ABA_n$. Through each point $A_1, A_2, \cdots, A_{n-1}$, construct $\overline{A_i B_i}$ parallel to $\overline{A_n B}$ that connect two sides of $\triangle AA_n B$.

DILATION METHOD: Construct a ray XY parallel to $\overline{AB}$. On the parallel ray, use a compass to mark off n equally spaced points $X_1, X_2, \cdots, X_n$ so that $XX_n \neq AB$. $\overleftrightarrow{AX}$ and $\overleftrightarrow{BX_n}$ intersect at a point O. Construct the rays OX_i that meet $\overline{AB}$ in points A_i.

Problem Set

1. Pretend you are the king or queen and that the length of your foot is the official measurement for one foot. Draw a line segment on a piece of paper that is the length of your foot. (You may have to remove your shoe.) Use the method above to find the length of 1 inch in your kingdom.

2. Using a ruler, draw a segment that is 10 cm. This length is referred to as a *decimeter*. Use the side splitter method to divide your segment into ten equal-sized pieces. What should be the length of each of those pieces based on your construction? Check the length of the pieces using a ruler. Are the lengths of the pieces accurate?

3. Repeat Problem 2 using the dilation method. What should be the length of each of those pieces based on your construction? Check the lengths of the pieces using a ruler. Are the lengths of the pieces accurate?

4. A portion of a ruler that measured whole centimeters is shown below. Determine the location of $5\frac{2}{3}$ cm on the portion of the ruler shown.

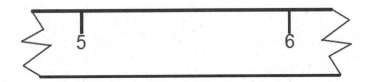

5. Merrick has a ruler that measures in inches only. He is measuring the length of a line segment that is between 8 in. and 9 in. Divide the one-inch section of Merrick's ruler below into eighths to help him measure the length of the segment.

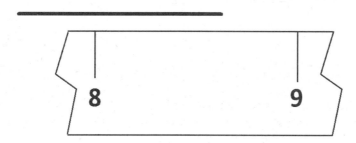

6. Use the dilation method to create an equally spaced 3 × 3 grid in the following square.

7. Use the side splitter method to create an equally spaced 3 × 3 grid in the following square.

This page intentionally left blank

Lesson 11: Dilations from Different Centers

Classwork

Exploratory Challenge 1

Drawing 2 and Drawing 3 are both scale drawings of Drawing 1.

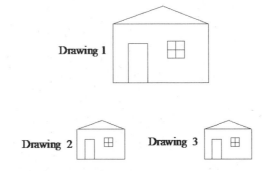

a. Determine the scale factor and center for each scale drawing. Take measurements as needed.

b. Is there a way to map Drawing 2 onto Drawing 3 or map Drawing 3 onto Drawing 2?

c. Generalize the parameters of this example and its results.

Exercise 1

Triangle ABC has been dilated with scale factor $\frac{1}{2}$ from centers O_1 and O_2. What can you say about line segments A_1A_2, B_1B_2, and C_1C_2?

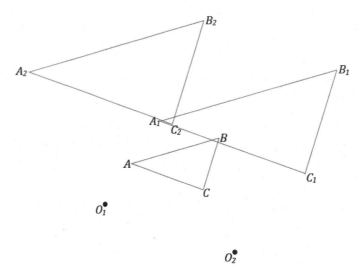

Exploratory Challenge 2

If Drawing 2 is a scale drawing of Drawing 1 with scale factor r_1 and Drawing 3 is a scale drawing of Drawing 2 with scale factor r_2, what is the relationship between Drawing 3 and Drawing 1?

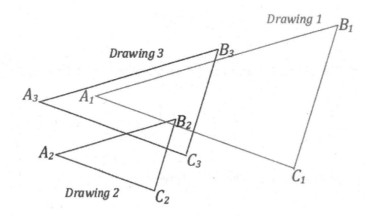

a. Determine the scale factor and center for each scale drawing. Take measurements as needed.

b. What is the scale factor going from Drawing 1 to Drawing 3? Take measurements as needed.

c. Compare the centers of dilations of Drawing 1 (to Drawing 2) and of Drawing 2 (to Drawing 3). What do you notice about these centers relative to the center of the composition of dilations O_3?

d. Generalize the parameters of this example and its results.

Exercise 2

Triangle ABC has been dilated with scale factor $\frac{2}{3}$ from center O_1 to get triangle $A'B'C'$, and then triangle $A'B'C'$ is dilated from center O_2 with scale factor $\frac{1}{2}$ to get triangle $A''B''C''$. Describe the dilation that maps triangle ABC to triangle $A''B''C''$. Find the center and scale factor for that dilation.

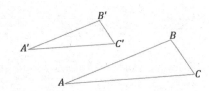

Lesson 11: Dilations from Different Centers

Lesson 11 M2
GEOMETRY

> **Lesson Summary**
>
> In a series of dilations, the scale factor that maps the original figure onto the final image is the product of all the scale factors in the series of dilations.

Problem Set

1. In Lesson 7, the dilation theorem for line segments said that if two different-length line segments in the plane were parallel to each other, then a dilation exists mapping one segment onto the other. Explain why the line segments must be different lengths for a dilation to exist.

2. Regular hexagon $A'B'C'D'E'F'$ is the image of regular hexagon $ABCDEF$ under a dilation from center O_1, and regular hexagon $A''B''C''D''E''F''$ is the image of regular hexagon $ABCDEF$ under a dilation from center O_2. Points A', B', C', D', E', and F' are also the images of points A'', B'', C'', D'', E'', and F'', respectively, under a translation along vector $\overrightarrow{D''D'}$. Find a possible regular hexagon $ABCDEF$.

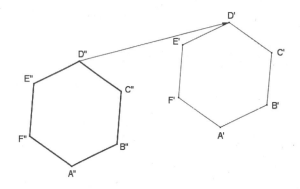

3. A dilation with center O_1 and scale factor $\frac{1}{2}$ maps figure F to figure F'. A dilation with center O_2 and scale factor $\frac{3}{2}$ maps figure F' to figure F''. Draw figures F' and F'', and then find the center O and scale factor r of the dilation that takes F to F''.

$O_1 \bullet \quad \bullet O_2$

4. A figure T is dilated from center O_1 with a scale factor $r_1 = \frac{3}{4}$ to yield image T', and figure T' is then dilated from center O_2 with a scale factor $r_2 = \frac{4}{3}$ to yield figure T''. Explain why $T \cong T''$.

5. A dilation with center O_1 and scale factor $\frac{1}{2}$ maps figure H to figure H'. A dilation with center O_2 and scale factor 2 maps figure H' to figure H''. Draw figures H' and H''. Find a vector for a translation that maps H to H''.

$\overset{\bullet}{O_1} \qquad \overset{\bullet}{O_2}$

Lesson 11: Dilations from Different Centers

6. Figure W is dilated from O_1 with a scale factor $r_1 = 2$ to yield W'. Figure W' is then dilated from center O_2 with a scale factor $r_2 = \frac{1}{4}$ to yield W''.

a. Construct the composition of dilations of figure W described above.

b. If you were to dilate figure W'', what scale factor would be required to yield an image that is congruent to figure W?

c. Locate the center of dilation that maps W'' to W using the scale factor that you identified in part (b).

7. Figures F_1 and F_2 in the diagram below are dilations of F from centers O_1 and O_2, respectively.

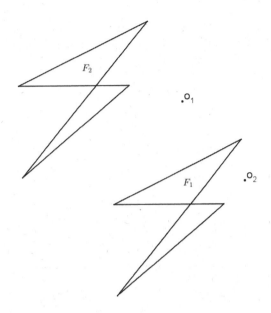

a. Find F.
b. If $F_1 \cong F_2$, what must be true of the scale factors r_1 and r_2 of each dilation?
c. Use direct measurement to determine each scale factor for D_{O_1, r_1} and D_{O_2, r_2}.

8. Use a coordinate plane to complete each part below using $U(2,3)$, $V(6,6)$, and $W(6,-1)$.
a. Dilate $\triangle UVW$ from the origin with a scale factor $r_1 = 2$. List the coordinates of image points U', V', and W'.
b. Dilate $\triangle UVW$ from $(0,6)$ with a scale factor of $r_2 = \frac{3}{4}$. List the coordinates of image points U'', V'', and W''.
c. Find the scale factor, r_3, from $\triangle U'V'W'$ to $\triangle U''V''W''$.
d. Find the coordinates of the center of dilation that maps $\triangle U'V'W'$ to $\triangle U''V''W''$.

This page intentionally left blank

A STORY OF FUNCTIONS

Lesson 12: What Are Similarity Transformations, and Why Do We Need Them?

Classwork

Opening Exercise

Observe Figures 1 and 2 and the images of the intermediate figures between them. Figures 1 and 2 are called *similar*.

What observations can we make about Figures 1 and 2?

Figure 1   Figure 2

? J

Definition:

A _____ _____ (or _____) is a composition of a finite number of dilations or basic rigid motions. The *scale factor* of a similarity transformation is the product of the scale factors of the dilations in the composition. If there are no dilations in the composition, the scale factor is defined to be 1.

Definition:

Two figures in a plane are _____ if there exists a similarity transformation taking one figure onto the other figure.

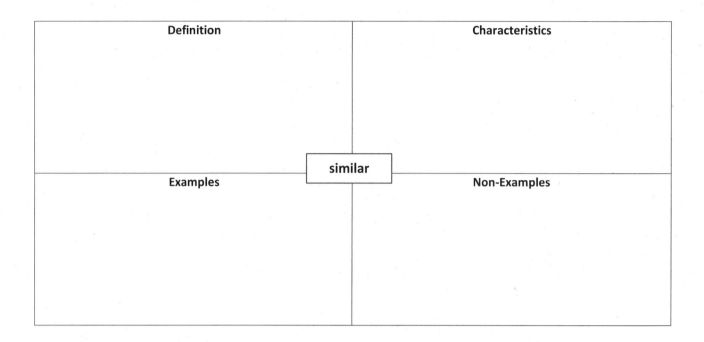

Example 1

Figure Z' is similar to Figure Z. Describe a transformation that maps Figure Z onto Figure Z'.

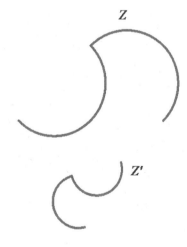

Exercises 1–3

1. Figure 1 is similar to Figure 2. Which transformations compose the similarity transformation that maps Figure 1 onto Figure 2?

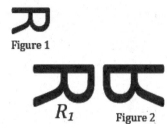

2. Figure S is similar to Figure S'. Which transformations compose the similarity transformation that maps S onto S'?

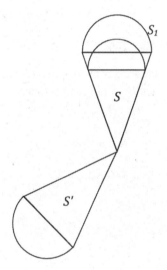

3. Figure 1 is similar to Figure 2. Which transformations compose the similarity transformation that maps Figure 1 onto Figure 2?

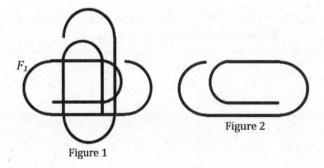

Lesson 12: What Are Similarity Transformations, and Why Do We Need Them?

Example 2

Show that no sequence of basic rigid motions and dilations takes the small figure to the large figure. Take measurements as needed.

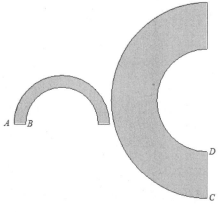

Exercises 4–5

4. Is there a sequence of dilations and basic rigid motions that takes the large figure to the small figure? Take measurements as needed.

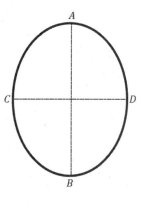
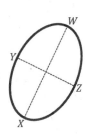

5. What purpose do transformations serve? Compare and contrast the application of rigid motions to the application of similarity transformations.

A STORY OF FUNCTIONS   Lesson 12   M2

GEOMETRY

> **Lesson Summary**
>
> Two figures are similar if there exists a similarity transformation that maps one figure onto the other.
>
> A similarity transformation is a composition of a finite number of dilations or rigid motions.

Problem Set

1. What is the relationship between scale drawings, dilations, and similar figures?
    a. How are scale drawings and dilations alike?
    b. How can scale drawings and dilations differ?
    c. What is the relationship of similar figures to scale drawings and dilations?

2. Given the diagram below, identify a similarity transformation, if one exists, that maps Figure A onto Figure B. If one does not exist, explain why.

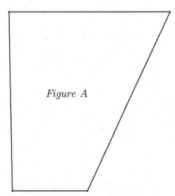

3. Teddy correctly identified a similarity transformation with at least one dilation that maps Figure I onto Figure II. Megan correctly identified a congruence transformation that maps Figure I onto Figure II. What must be true about Teddy's similarity transformation?

Lesson 12:   What Are Similarity Transformations, and Why Do We Need Them?   S.91

4. Given the coordinate plane shown, identify a similarity transformation, if one exists, that maps X onto Y. If one does not exist, explain why.

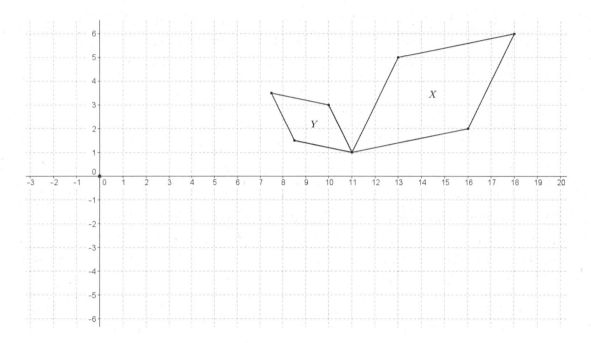

5. Given the diagram below, identify a similarity transformation, if one exists, that maps G onto H. If one does not exist, explain why. Provide any necessary measurements to justify your answer.

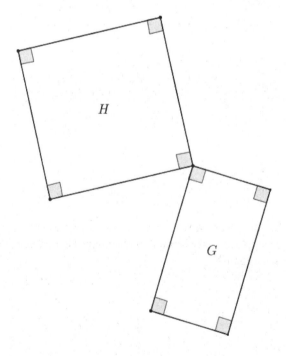

6. Given the coordinate plane shown, identify a similarity transformation, if one exists, that maps $ABCD$ onto $A'''B'''C'''D'''$. If one does not exist, explain why.

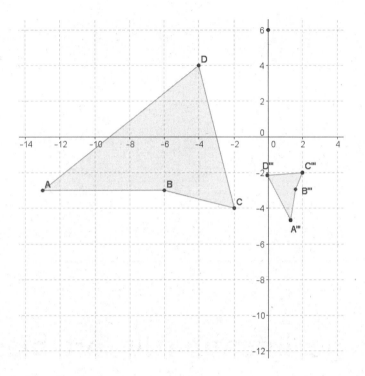

7. The diagram below shows a dilation of the plane ... or does it? Explain your answer.

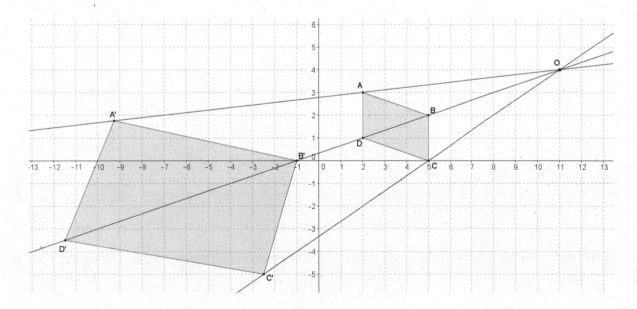

This page intentionally left blank

Lesson 13: Properties of Similarity Transformations

Classwork

Example 1

Similarity transformation G consists of a rotation about the point P by $90°$, followed by a dilation centered at P with a scale factor of $r = 2$, and then followed by a reflection across line ℓ. Find the image of the triangle.

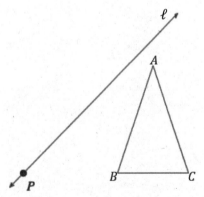

Example 2

A similarity transformation G applied to trapezoid $ABCD$ consists of a translation by vector $\overrightarrow{XY}$, followed by a reflection across line m, and then followed by a dilation centered at P with a scale factor of $r = 2$. Recall that we can describe the same sequence using the following notation: $D_{P,2}\left(r_m\left(T_{XY}(ABCD)\right)\right)$. Find the image of $ABCD$.

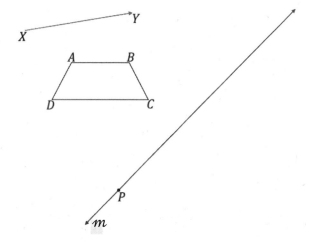

Lesson 13: Properties of Similarity Transformations

Exercise 1

A similarity transformation for triangle DEF is described by $r_n\left(D_{A,\frac{1}{2}}\left(R_{A,90°}(DEF)\right)\right)$. Locate and label the image of triangle DEF under the similarity.

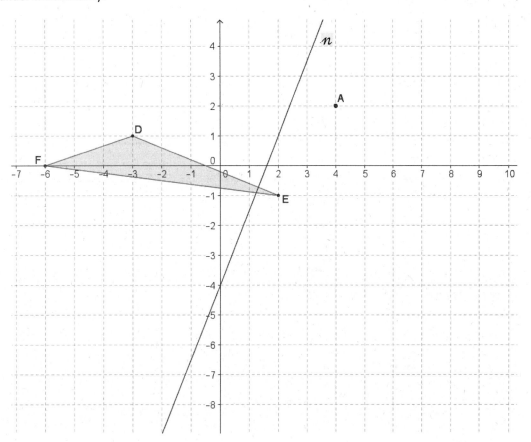

Lesson Summary

Properties of similarity transformations:

1. Distinct points are mapped to distinct points.
2. Each point P' in the plane has a pre-image.
3. There is a scale factor of r for G so that for any pair of points P and Q with images $P' = G(P)$ and $Q' = G(Q)$, then $P'Q' = rPQ$.
4. A similarity transformation sends lines to lines, rays to rays, line segments to line segments, and parallel lines to parallel lines.
5. A similarity transformation sends angles to angles of equal measure.
6. A similarity transformation maps a circle of radius R to a circle of radius rR, where r is the scale factor of the similarity transformation.

Problem Set

1. A similarity transformation consists of a reflection over line ℓ, followed by a dilation from O with a scale factor of $r = \frac{3}{4}$. Use construction tools to find $\triangle G''H''I''$.

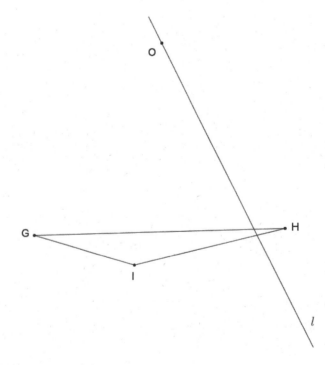

2. A similarity transformation consists of a dilation from point O with a scale factor of $r = 2\frac{1}{2}$, followed by a rotation about O of $-90°$. Use construction tools to find kite $A''B''C''D''$.

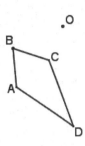

3. For the Figure Z, find the image of $r_\ell(R_{P,90°}(D_{P,\frac{1}{2}}(Z)))$.

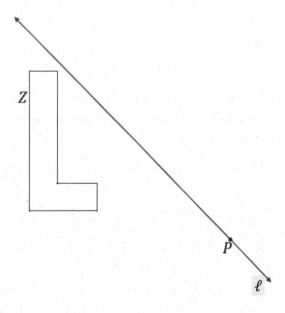

Lesson 13: Properties of Similarity Transformations

4. A similarity transformation consists of a translation along vector $\overrightarrow{UV}$, followed by a rotation of $60°$ about P, then followed by a dilation from P with a scale factor of $r = \frac{1}{3}$. Use construction tools to find $\triangle X'''Y'''Z'''$.

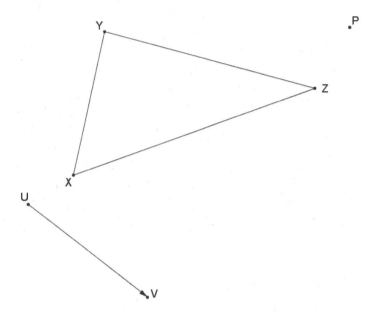

5. Given the quarter-circular figure determined by points A, B, and C, a similarity transformation consists of a $-65°$ rotation about point B, followed by a dilation from point O with a scale factor of $r = \frac{1}{2}$. Find the image of the figure determined by points A'', B'', and C''.

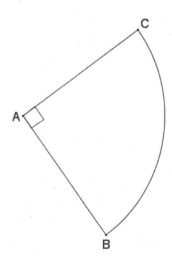

Describe a different similarity transformation that would map quarter-circle ABC to quarter-circle $A''B''C''$.

6. A similarity transformation consists of a dilation from center O with a scale factor of $\frac{1}{2}$, followed by a rotation of 60° about point O. Complete the similarity transformation on Figure T to complete the drawing of Figure T''.

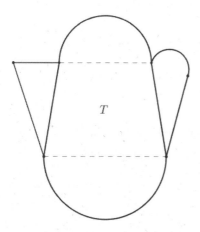

7. Given Figure R on the coordinate plane shown below, a similarity transformation consists of a dilation from $(0,6)$ with a scale factor of $\frac{1}{4}$, followed by a reflection over line $x = -1$, and then followed by a vertical translation of 5 units down. Find the image of Figure R.

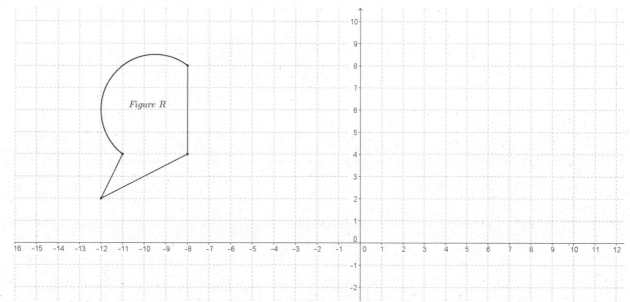

Lesson 13: Properties of Similarity Transformations

A STORY OF FUNCTIONS   Lesson 13   M2

GEOMETRY

8. Given $\triangle ABC$, with vertices $A(2,-7)$, $B(-2,-1)$, and $C(3,-4)$, locate and label the image of the triangle under the similarity transformation $D_{B',\frac{1}{2}}\left(R_{A,120°}\left(r_{x=2}(ABC)\right)\right)$.

9. In Problem 8, describe the relationship of A''' to $\overline{AB'}$, and explain your reasoning.

10. Given $O(-8,3)$ and quadrilateral $BCDE$, with $B(-5,1)$, $C(-6,-1)$, $D(-4,-1)$, and $E(-4,2)$, what are the coordinates of the vertices of the image of $BCDE$ under the similarity transformation $r_{x-axis}\left(D_{O,3}(BCDE)\right)$?

11. Given triangle ABC as shown on the diagram of the coordinate plane:

    a. Perform a translation so that vertex A maps to the origin.
    b. Next, dilate the image $A'B'C'$ from the origin using a scale factor of $\frac{1}{3}$.
    c. Finally, translate the image $A''B''C''$ so that the vertex A'' maps to the original point A.
    d. Using transformations, describe how the resulting image $A'''B''C''$ relates to the original figure ABC.

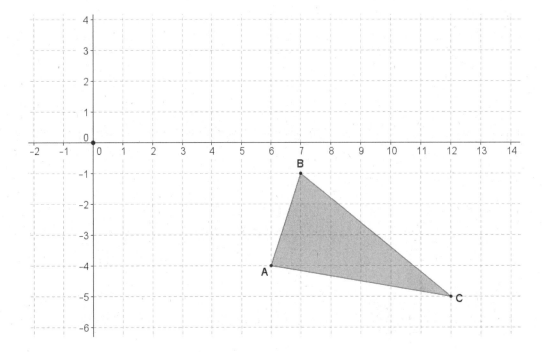

S.102   Lesson 13:   Properties of Similarity Transformations

12.
   a. In the coordinate plane, name the single transformation resulting from the composition of the two dilations shown below:

   $D_{(0,0),2}$ followed by $D_{(0,4),\frac{1}{2}}$

   (Hint: Try it!)

   b. In the coordinate plane, name the single transformation resulting from the composition of the two dilations shown below:

   $D_{(0,0),2}$ followed by $D_{(4,4),\frac{1}{2}}$

   (Hint: Try it!)

   c. Using the results from parts (a) and (b), describe what happens to the origin under both similarity transformations.

This page intentionally left blank

Example 1

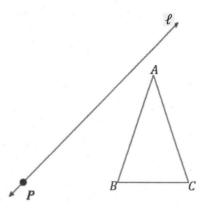

Example 2

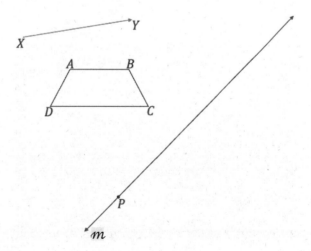

This page intentionally left blank

Lesson 14: Similarity

Classwork

Example 1

We said that for a figure A in the plane, it must be true that $A \sim A$. Describe why this must be true.

Example 2

We said that for figures A and B in the plane so that $A \sim B$, then it must be true that $B \sim A$. Describe why this must be true.

Example 3

Based on the definition of *similar,* how would you show that any two circles are similar?

A STORY OF FUNCTIONS   Lesson 14   M2

GEOMETRY

Example 4

Suppose $\triangle ABC \leftrightarrow \triangle DEF$ and that, under this correspondence, corresponding angles are equal and corresponding sides are proportional. Does this guarantee that $\triangle ABC$ and $\triangle DEF$ are similar?

Example 5

a. In the diagram below, $\triangle ABC \sim \triangle A'B'C'$. Describe a similarity transformation that maps $\triangle ABC$ to $\triangle A'B'C'$.

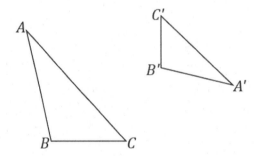

b. Joel says the sequence must require a dilation and three rigid motions, but Sharon is sure there is a similarity transformation composed of just a dilation and two rigid motions. Who is right?

S.108   Lesson 14:   Similarity

A STORY OF FUNCTIONS   Lesson 14   M2

GEOMETRY

Lesson Summary

Similarity is reflexive because a figure is similar to itself.

Similarity is symmetric because once a similarity transformation is determined to take a figure to another, there are inverse transformations that can take the figure back to the original.

Problem Set

1. If you are given any two congruent triangles, describe a sequence of basic rigid motions that takes one to the other.

2. If you are given two similar triangles that are not congruent triangles, describe a sequence of dilations and basic rigid motions that takes one to the other.

3. Given two line segments, $\overline{AB}$ and $\overline{CD}$, of different lengths, answer the following questions:
   a. It is always possible to find a similarity transformation that maps $\overline{AB}$ to $\overline{CD}$ sending A to C and B to D. Describe one such similarity transformation.
   b. If you are given that $\overline{AB}$ and $\overline{CD}$ are not parallel, are not congruent, do not share any points, and do not lie in the same line, what is the fewest number of transformations needed in a sequence to map $\overline{AB}$ to $\overline{CD}$? Which transformations make this work?
   c. If you performed a similarity transformation that instead takes A to D and B to C, either describe what mistake was made in the similarity transformation, or describe what additional transformation is needed to fix the error so that A maps to C and B maps to D.

4. We claim that similarity is transitive (i.e., if A, B, and C are figures in the plane such that $A \sim B$ and $B \sim C$, then $A \sim C$). Describe why this must be true.

5. Given two line segments, $\overline{AB}$ and $\overline{CD}$, of different lengths, we have seen that it is always possible to find a similarity transformation that maps $\overline{AB}$ to $\overline{CD}$, sending A to C and B to D with one rotation and one dilation. Can you do this with one reflection and one dilation?

6. Given two triangles, $\triangle ABC \sim \triangle DEF$, is it always possible to rotate $\triangle ABC$ so that the sides of $\triangle ABC$ are parallel to the corresponding sides in $\triangle DEF$ (e.g., $\overline{AB} \parallel \overline{DE}$)?

Lesson 14:   Similarity   S.109

This page intentionally left blank

Lesson 15: The Angle-Angle (AA) Criterion for Two Triangles to Be Similar

Classwork

Exercises

1. Draw two triangles of different sizes with two pairs of equal angles. Then, measure the lengths of the corresponding sides to verify that the ratio of their lengths is proportional. Use a ruler, compass, or protractor, as necessary.

2. Are the triangles you drew in Exercise 1 similar? Explain.

3. Why is it that you only need to construct triangles where two pairs of angles are equal but not three?

4. Why were the ratios of the corresponding sides proportional?

5. Do you think that what you observed will be true when you construct a pair of triangles with two pairs of equal angles? Explain.

6. Draw another two triangles of different sizes with two pairs of equal angles. Then, measure the lengths of the corresponding sides to verify that the ratio of their lengths is proportional. Use a ruler, compass, or protractor, as necessary.

7. Are the triangles shown below similar? Explain. If the triangles are similar, identify any missing angle and side-length measures.

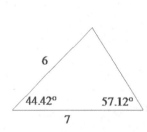
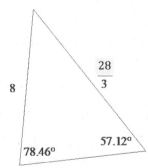

8. Are the triangles shown below similar? Explain. If the triangles are similar, identify any missing angle and side-length measures.

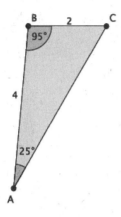
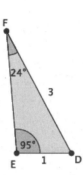

9. The triangles shown below are similar. Use what you know about similar triangles to find the missing side lengths x and y.

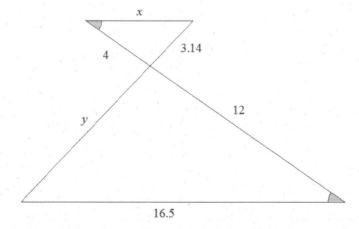

10. The triangles shown below are similar. Write an explanation to a student, Claudia, of how to find the lengths of x and y.

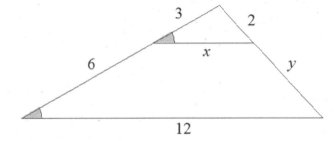

Problem Set

1. In the figure to the right, △ LMN ~ △ MPL.

   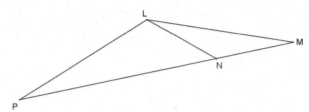

   a. Classify △ LMP based on what you know about similar triangles, and justify your reasoning.
   b. If $m\angle P = 20°$, find the remaining angles in the diagram.

2. In the diagram below, △ ABC ~ △ AFD. Determine whether the following statements must be true from the given information, and explain why.

   a. △ CAB ~ △ DAF
   b. △ ADF ~ △ CAB
   c. △ BCA ~ △ ADF
   d. △ ADF ~ △ ACB

   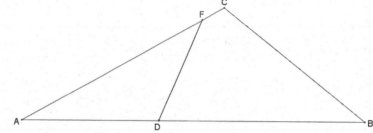

3. In the diagram below, D is the midpoint of $\overline{AB}$, F is the midpoint of $\overline{BC}$, and E is the midpoint of $\overline{AC}$. Prove that △ ABC ~ △ FED.

   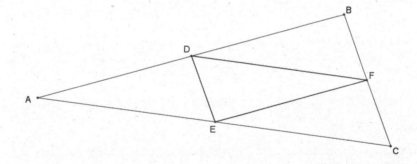

4. Use the diagram below to answer each part.

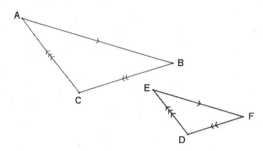

a. If $\overline{AC} \parallel \overline{ED}$, $\overline{AB} \parallel \overline{EF}$, and $\overline{CB} \parallel \overline{DF}$, prove that the triangles are similar.

b. The triangles are not congruent. Find the dilation that takes one to the other.

5. Given trapezoid $ABDE$, and $\overline{AB} \parallel \overline{ED}$, prove that $\triangle AFB \sim \triangle DEF$.

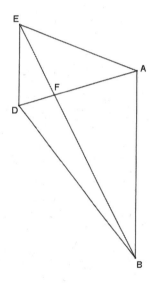

Cutouts to use for in-class discussion:

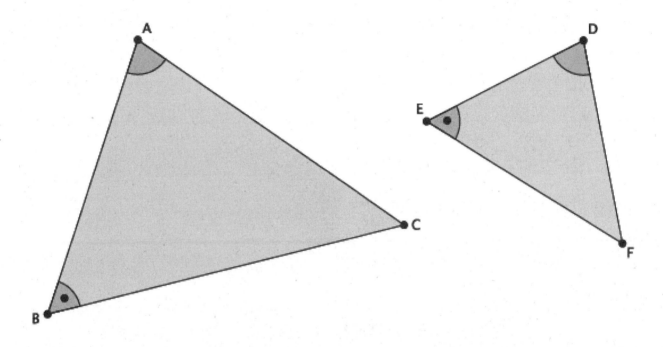

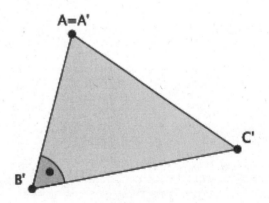

This page intentionally left blank

Lesson 16: Between-Figure and Within-Figure Ratios

Classwork

Opening Exercise

At a certain time of day, a 12 m flagpole casts an 8 m shadow. Write an equation that would allow you to find the height, h, of the tree that uses the length, s, of the tree's shadow.

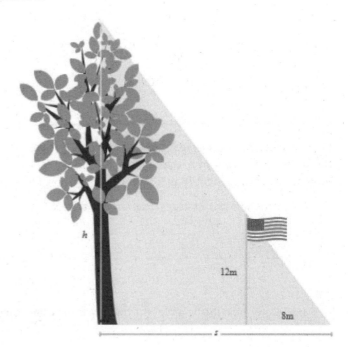

Example 1

Given $\triangle ABC \sim \triangle A'B'C'$, find the missing side lengths.

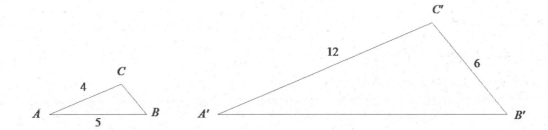

Example 2

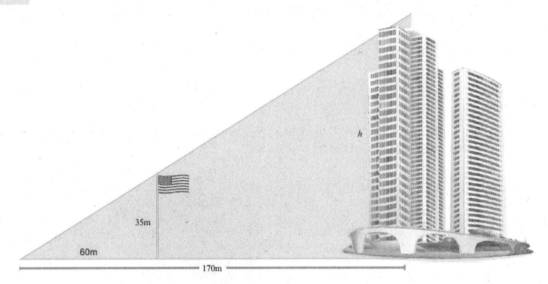

In the diagram above, a large flagpole stands outside of an office building. Marquis realizes that when he looks up from the ground 60 m away from the flagpole, the top of the flagpole and the top of the building line up. If the flagpole is 35 m tall and Marquis is 170 m from the building, how tall is the building?

a. Are the triangles in the diagram similar? Explain.

b. Determine the height of the building using what you know about scale factors.

c. Determine the height of the building using ratios *between* similar figures.

d. Determine the height of the building using ratios *within* similar figures.

Lesson 16: Between-Figure and Within-Figure Ratios

Example 3

The following right triangles are similar. We will determine the unknown side lengths by using ratios within the first triangle. For each of the triangles below, we define the base as the horizontal length of the triangle and the height as the vertical length.

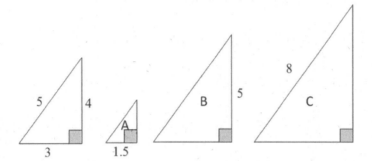

a. Write and find the value of the ratio that compares the height to the hypotenuse of the leftmost triangle.

b. Write and find the value of the ratio that compares the base to the hypotenuse of the leftmost triangle.

c. Write and find the value of the ratio that compares the height to the base of the leftmost triangle.

d. Use the triangle with lengths 3– 4– 5 and triangle A to answer the following questions:
   i. Which ratio can be used to determine the height of triangle A?

   ii. Which ratio can be used to determine the hypotenuse of triangle A?

   iii. Find the unknown lengths of triangle A.

Lesson 16: Between-Figure and Within-Figure Ratios

e. Use the triangle with lengths 3–4–5 and triangle B to answer the following questions:
   i. Which ratio can be used to determine the base of triangle B?

   ii. Which ratio can be used to determine the hypotenuse of triangle B?

   iii. Find the unknown lengths of triangle B.

f. Use the triangle with lengths 3–4–5 and triangle C to answer the following questions:
   i. Which ratio can be used to determine the height of triangle C?

   ii. Which ratio can be used to determine the base of triangle C?

   iii. Find the unknown lengths of triangle C.

g. Explain the relationship of the ratio of the corresponding sides within a figure to the ratio of the corresponding sides within a similar figure.

h. How does the relationship you noted in part (g) allow you to determine the length of an unknown side of a triangle?

A STORY OF FUNCTIONS Lesson 16 M2
GEOMETRY

Problem Set

1. △ DEF ~ △ ABC All side length measurements are in centimeters. Use between-figure ratios and/or within-figure ratios to determine the unknown side lengths.

2. Given △ ABC ~ △ XYZ, answer the following questions:
   a. Write and find the value of the ratio that compares the height $\overline{AC}$ to the hypotenuse of △ ABC.
   b. Write and find the value of the ratio that compares the base $\overline{AB}$ to the hypotenuse of △ ABC.
   c. Write and find the value of the ratio that compares the height $\overline{AC}$ to the base $\overline{AB}$ of △ ABC.
   d. Use within-figure ratios to find the corresponding height of △ XYZ.
   e. Use within-figure ratios to find the hypotenuse of △ XYZ.

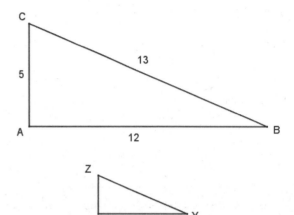

3. Right triangles A, B, C, and D are similar. Determine the unknown side lengths of each triangle by using ratios of side lengths within triangle A.

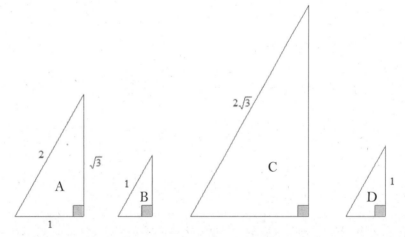

   a. Write and find the value of the ratio that compares the height to the hypotenuse of triangle A.
   b. Write and find the value of the ratio that compares the base to the hypotenuse of triangle A.
   c. Write and find the value of the ratio that compares the height to the base of triangle A.
   d. Which ratio can be used to determine the height of triangle B? Find the height of triangle B.
   e. Which ratio can be used to determine the base of triangle B? Find the base of triangle B.
   f. Find the unknown lengths of triangle C.
   g. Find the unknown lengths of triangle D.

Lesson 16: Between-Figure and Within-Figure Ratios

h. Triangle E is also similar to triangles A, B, C, and D. Find the lengths of the missing sides in terms of x.

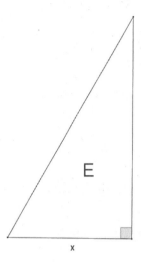

4. Brian is photographing the Washington Monument and wonders how tall the monument is. Brian places his 5 ft. camera tripod approximately 100 yd. from the base of the monument. Lying on the ground, he visually aligns the top of his tripod with the top of the monument and marks his location on the ground approximately 2 ft. 9 in. from the center of his tripod. Use Brian's measurements to approximate the height of the Washington Monument.

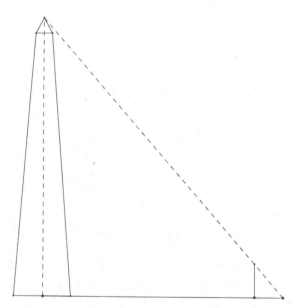

5. Catarina's boat has come untied and floated away on the lake. She is standing atop a cliff that is 35 ft. above the water in a lake. If she stands 10 ft. from the edge of the cliff, she can visually align the top of the cliff with the water at the back of her boat. Her eye level is $5\frac{1}{2}$ ft. above the ground. Approximately how far out from the cliff is Catarina's boat?

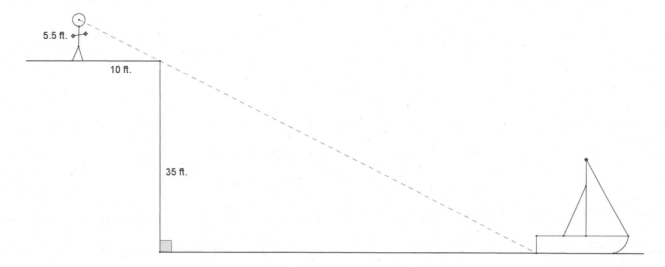

6. Given the diagram below and $\triangle ABC \sim \triangle XYZ$, find the unknown lengths x, $2x$, and $3x$.

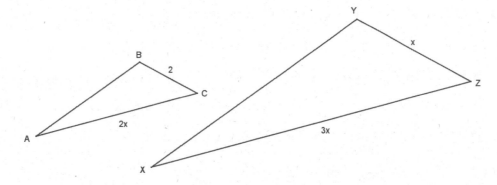

This page intentionally left blank

Lesson 17: The Side-Angle-Side (SAS) and Side-Side-Side (SSS) Criteria for Two Triangles to Be Similar

Classwork

Opening Exercise

a. Choose three lengths that represent the sides of a triangle. Draw the triangle with your chosen lengths using construction tools.

b. Multiply each length in your original triangle by 2 to get three corresponding lengths of sides for a second triangle. Draw your second triangle using construction tools.

c. Do your constructed triangles appear to be similar? Explain your answer.

d. Do you think that the triangles can be shown to be similar without knowing the angle measures?

Exploratory Challenge 1/Exercises 1–2

1. Examine the figure, and answer the questions to determine whether or not the triangles shown are similar.

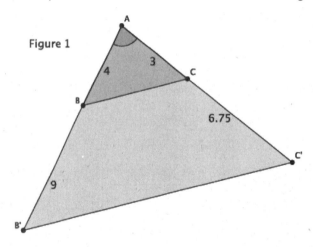

Figure 1

a. What information is given about the triangles in Figure 1?

b. How can the information provided be used to determine whether $\triangle ABC$ is similar to $\triangle AB'C'$?

c. Compare the corresponding side lengths of △ABC and △AB'C'. What do you notice?

d. Based on your work in parts (a)–(c), draw a conclusion about the relationship between △ABC and △AB'C'. Explain your reasoning.

2. Examine the figure, and answer the questions to determine whether or not the triangles shown are similar.

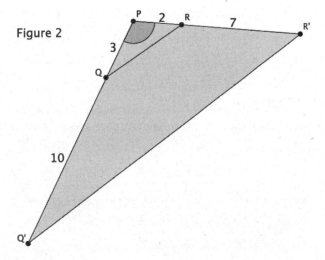

a. What information is given about the triangles in Figure 2?

b. How can the information provided be used to determine whether △ PQR is similar to △ $PQ'R'$?

c. Compare the corresponding side lengths of △ PQR and △ $PQ'R'$. What do you notice?

d. Based on your work in parts (a)–(c), draw a conclusion about the relationship between △ PQR and △ $PQ'R'$. Explain your reasoning.

Exploratory Challenge 2/Exercises 3–4

3. Examine the figure, and answer the questions to determine whether or not the triangles shown are similar.

Figure 3

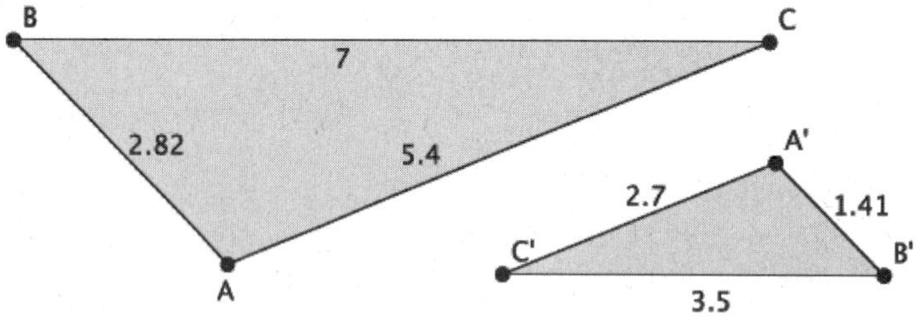

a. What information is given about the triangles in Figure 3?

A STORY OF FUNCTIONS   Lesson 17   M2

GEOMETRY

b. How can the information provided be used to determine whether △ ABC is similar to △ AB'C'?

c. Compare the corresponding side lengths of △ ABC and △ AB'C'. What do you notice?

d. Based on your work in parts (a)–(c), make a conjecture about the relationship between △ ABC and △ AB'C'. Explain your reasoning.

4. Examine the figure, and answer the questions to determine whether or not the triangles shown are similar.

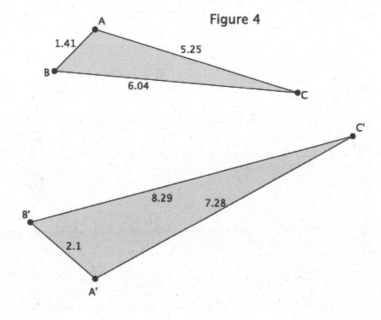

Figure 4

Lesson 17:   The Side-Angle-Side (SAS) and Side-Side-Side (SSS) Criteria for Two Triangles to Be Similar

S.131

a. What information is given about the triangles in Figure 4?

b. How can the information provided be used to determine whether △ABC is similar to △AB'C'?

c. Compare the corresponding side lengths of △ABC and △AB'C'. What do you notice?

d. Based on your work in parts (a)–(c), make a conjecture about the relationship between △ABC and △AB'C'. Explain your reasoning.

Exercises 5–10

5. Are the triangles shown below similar? Explain. If the triangles are similar, write the similarity statement.

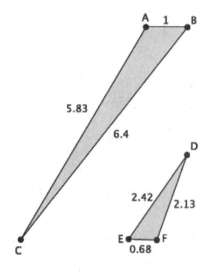

6. Are the triangles shown below similar? Explain. If the triangles are similar, write the similarity statement.

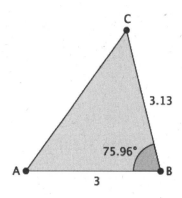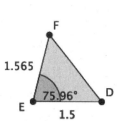

7. Are the triangles shown below similar? Explain. If the triangles are similar, write the similarity statement.

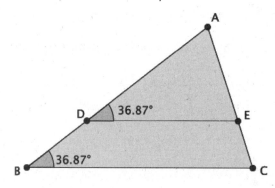

8. Are the triangles shown below similar? Explain. If the triangles are similar, write the similarity statement.

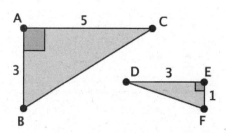

A STORY OF FUNCTIONS Lesson 17 M2
GEOMETRY

9. Are the triangles shown below similar? Explain. If the triangles are similar, write the similarity statement.

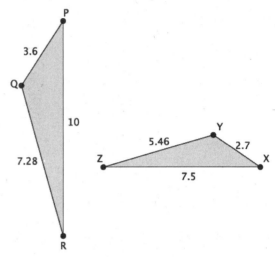

10. Are the triangles shown below similar? Explain. If the triangles are similar, write the similarity statement.

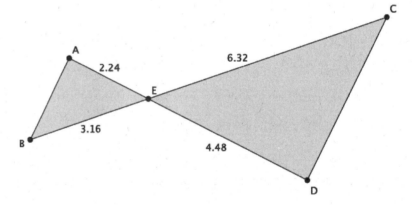

A STORY OF FUNCTIONS   Lesson 17   M2
GEOMETRY

Problem Set

1. For parts (a) through (d) below, state which of the three triangles, if any, are similar and why.

   a.

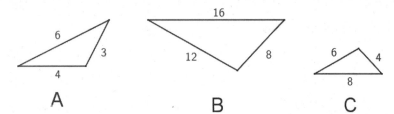

   b.

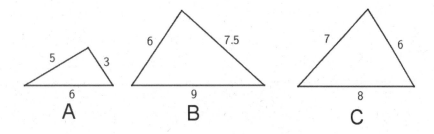

   c.

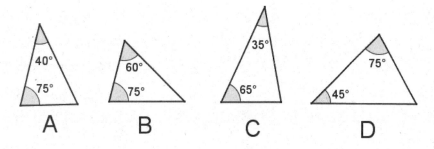

   d.

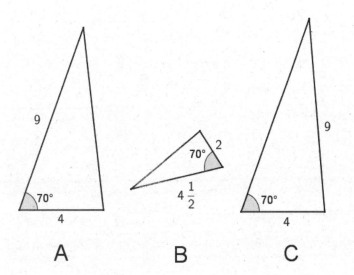

Lesson 17: The Side-Angle-Side (SAS) and Side-Side-Side (SSS) Criteria for Two Triangles to Be Similar

2. For each given pair of triangles, determine if the triangles are similar or not, and provide your reasoning. If the triangles are similar, write a similarity statement relating the triangles.

   a.
   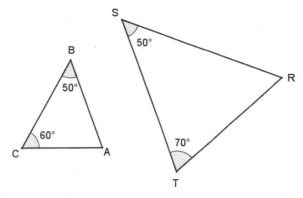

   b.
   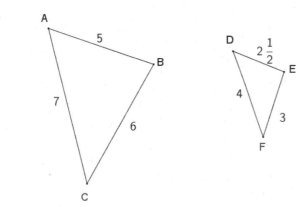

   c.
   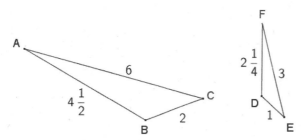

   d.
   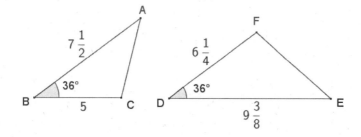

3. For each pair of similar triangles below, determine the unknown lengths of the sides labeled with letters.

   a.

   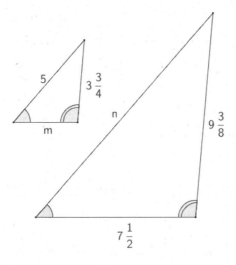

   b.

   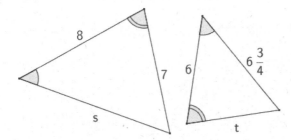

4. Given that $\overline{AD}$ and $\overline{BC}$ intersect at E and $\overline{AB} \parallel \overline{CD}$, show that $\triangle ABE \sim \triangle DCE$.

   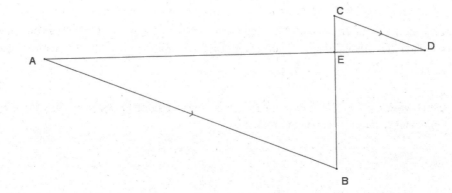

Lesson 17: The Side-Angle-Side (SAS) and Side-Side-Side (SSS) Criteria for Two Triangles to Be Similar

5. Given $BE = 11$, $EA = 11$, $BD = 7$, and $DC = 7$, show that $\triangle BED \sim \triangle BAC$.

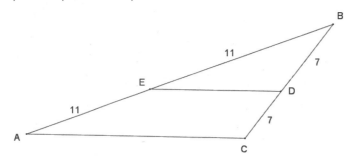

6. Given the diagram below, X is on $\overline{RS}$ and Y is on $\overline{RT}$, $XS = 2$, $XY = 6$, $ST = 9$, and $YT = 4$.

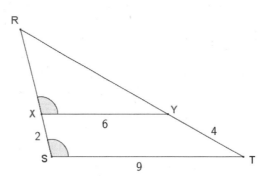

   a. Show that $\triangle RXY \sim \triangle RST$.
   b. Find RX and RY.

7. One triangle has a 120° angle, and a second triangle has a 65° angle. Is it possible that the two triangles are similar? Explain why or why not.

8. A right triangle has a leg that is 12 cm, and another right triangle has a leg that is 6 cm. Can you tell whether the two triangles are similar? If so, explain why. If not, what other information would be needed to show they are similar?

9. Given the diagram below, $JH = 7.5$, $HK = 6$, and $KL = 9$, is there a pair of similar triangles? If so, write a similarity statement, and explain why. If not, explain your reasoning.

Lesson 18: Similarity and the Angle Bisector Theorem

Classwork

Opening Exercise

a. What is an angle bisector?

b. Describe the angle relationships formed when parallel lines are cut by a transversal.

c. What are the properties of an isosceles triangle?

Discussion

In the diagram below, the angle bisector of $\angle A$ in $\triangle ABC$ meets side $\overline{BC}$ at point D. Does the angle bisector create any observable relationships with respect to the side lengths of the triangle?

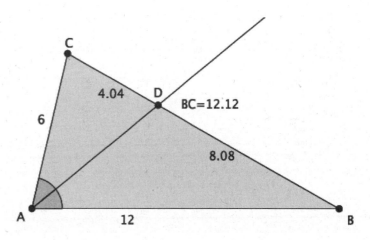

Exercises 1–4

1. The sides of a triangle are 8, 12, and 15. An angle bisector meets the side of length 15. Find the lengths x and y. Explain how you arrived at your answers.

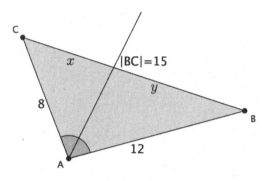

2. The sides of a triangle are 8, 12, and 15. An angle bisector meets the side of length 12. Find the lengths x and y.

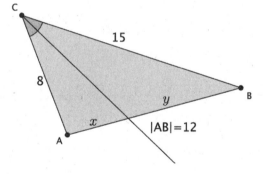

Lesson 18: Similarity and the Angle Bisector Theorem

3. The sides of a triangle are 8, 12, and 15. An angle bisector meets the side of length 8. Find the lengths x and y.

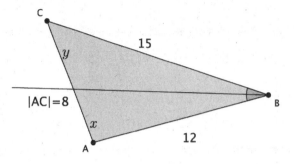

4. The angle bisector of an angle splits the opposite side of a triangle into lengths 5 and 6. The perimeter of the triangle is 33. Find the lengths of the other two sides.

Problem Set

1. The sides of a triangle have lengths of 5, 8, and $6\frac{1}{2}$. An angle bisector meets the side of length $6\frac{1}{2}$. Find the lengths x and y.

2. The sides of a triangle are $10\frac{1}{2}$, $16\frac{1}{2}$, and 9. An angle bisector meets the side of length 9. Find the lengths x and y.

3. In the diagram of triangle DEF below, $\overline{DG}$ is an angle bisector, $DE = 8$, $DF = 6$, and $EF = 8\frac{1}{6}$. Find FG and EG.

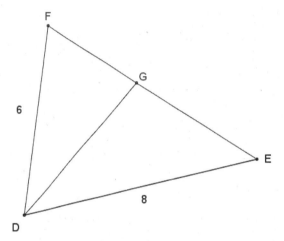

4. Given the diagram below and $\angle BAD \cong \angle DAC$, show that $BD:BA = CD:CA$.

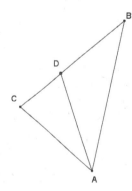

5. The perimeter of triangle LMN is 32 cm. $\overline{NX}$ is the angle bisector of angle N, $LX = 3$ cm, and $XM = 5$ cm. Find LN and MN.

6. Given $CD = 3$, $DB = 4$, $BF = 4$, $FE = 5$, $AB = 6$, and $\angle CAD \cong \angle DAB \cong \angle BAF \cong \angle FAE$, find the perimeter of quadrilateral $AEBC$.

7. If $\overline{AE}$ meets $\overline{BC}$ at D such that $CD:BD = CA:BA$, show that $\angle CAD \cong \angle BAD$. Explain how this proof relates to the angle bisector theorem.

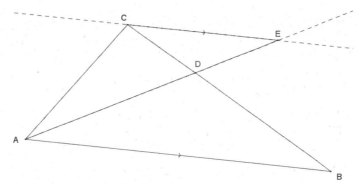

8. In the diagram below, $\overline{ED} \cong \overline{DB}$, $\overline{BE}$ bisects $\angle ABC$, $AD = 4$, and $DC = 8$. Prove that $\triangle ADB \sim \triangle CEB$.

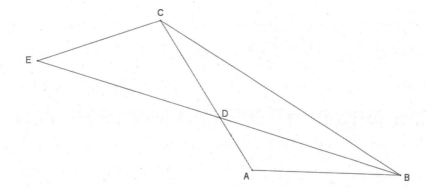

Lesson 18: Similarity and the Angle Bisector Theorem

This page intentionally left blank

Lesson 19: Families of Parallel Lines and the Circumference of the Earth

Classwork

Opening Exercise

Show $x:y = x':y'$ is equivalent to $x:x' = y:y'$.

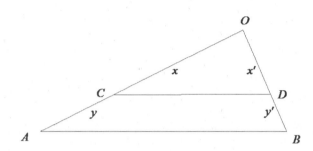

Exercises 1–2

Lines that appear to be parallel are in fact parallel.

1.

2.

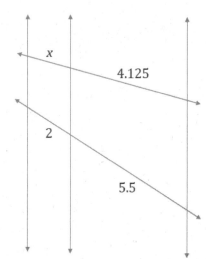

Problem Set

1. Given the diagram shown, $\overline{AD} \parallel \overline{GJ} \parallel \overline{LO} \parallel \overline{QT}$, and $\overline{AQ} \parallel \overline{BR} \parallel \overline{CS} \parallel \overline{DT}$. Use the additional information given in each part below to answer the questions:

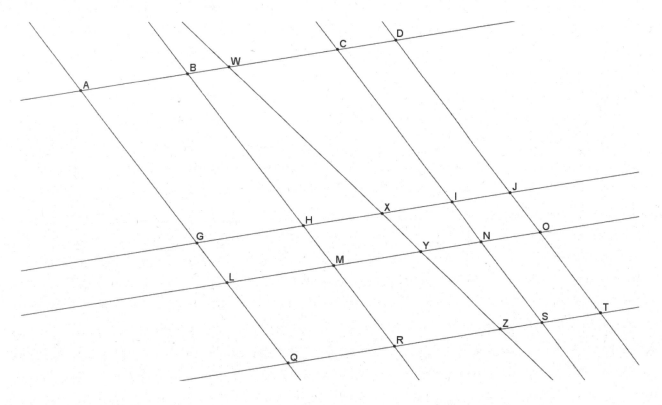

   a. If $GL = 4$, what is HM?
   b. If $GL = 4$, $LQ = 9$, and $XY = 5$, what is YZ?
   c. Using information from part (b), if $CI = 18$, what is WX?

2. Use your knowledge about families of parallel lines to find the coordinates of point P on the coordinate plane below.

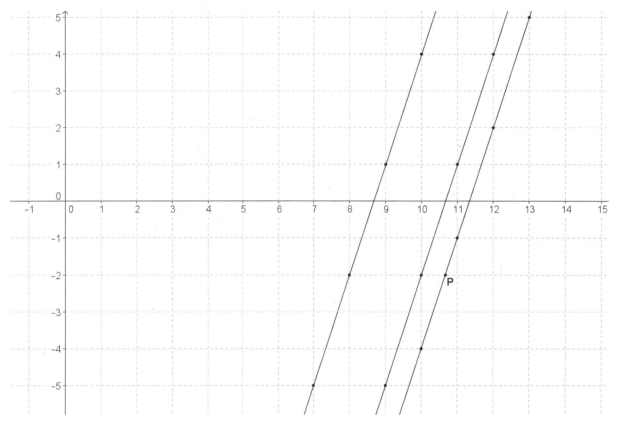

3. $ACDB$ and $FCDE$ are both trapezoids with bases $\overline{AB}$, $\overline{FE}$, and $\overline{CD}$. The perimeter of trapezoid $ACDB$ is $24\frac{1}{2}$. If the ratio of $AF:FC$ is $1:3$, $AB = 7$, and $ED = 5\frac{5}{8}$, find AF, FC, and BE.

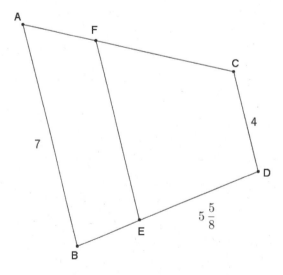

4. Given the diagram and the ratio of $a:b$ is $3:2$, answer each question below:

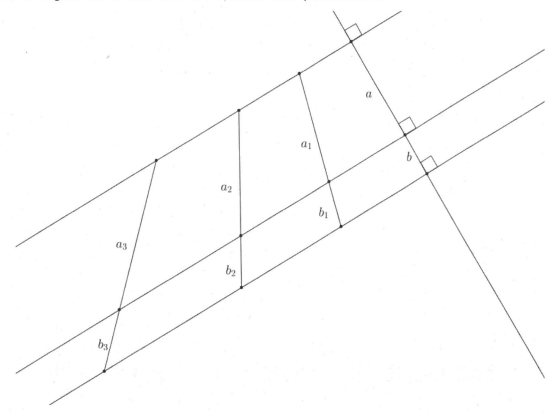

a. Write an equation for a_n in terms of b_n.
b. Write an equation for b_n in terms of a_n.
c. Use one of your equations to find b_1 in terms of a if $a_1 = 1.2(a)$.
d. What is the relationship between b_1 and b?
e. What constant, c, relates b_1 and b? Is this surprising? Why or why not?
f. Using the formula $a_n = c \cdot a_{n-1}$, find a_3 in terms of a.
g. Using the formula $b_n = c \cdot b_{n-1}$, find b_3 in terms of b.
h. Use your answers from parts (f) and (g) to calculate the value of the ratio of $a_3 : b_3$.

5. Julius wants to try to estimate the circumference of the earth based on measurements made near his home. He cannot find a location near his home where the sun is straight overhead. Will he be able to calculate the circumference of the earth? If so, explain and draw a diagram to support your claim.

This page intentionally left blank

Lesson 20: How Far Away Is the Moon?

Classwork

Opening Exercise

What is a solar eclipse? What is a lunar eclipse?

Discussion

Solar Eclipse

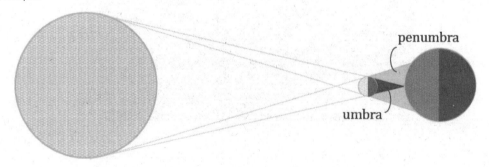

3D view:

Lunar Eclipse

3D view:

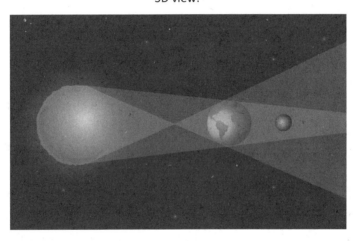

Example

a. If the circumference of the earth is about 25,000 miles, what is the earth's diameter in miles?

Lesson 20: How Far Away Is the Moon?

b. Using part (a), what is the moon's diameter in miles?

c. How far away is the moon in miles?

Problem Set

1. If the sun and the moon do not have the same diameter, explain how the sun's light can be covered by the moon during a solar eclipse.

2. What would a lunar eclipse look like when viewed from the moon?

3. Suppose you live on a planet with a moon, where during a solar eclipse, the moon appears to be half the diameter of the sun.

   a. Draw a diagram of how the moon would look against the sun during a solar eclipse.

   b. A 1-inch diameter marble held 100 inches away on the planet barely blocks the sun. How many moon diameters away is the moon from the planet? Draw and label a diagram to support your answer.

   c. If the diameter of the moon is approximately $\frac{3}{5}$ of the diameter of the planet and the circumference of the planet is 185,000 miles, approximately how far is the moon from the planet?

Lesson 21: Special Relationships Within Right Triangles—Dividing into Two Similar Sub-Triangles

Classwork

Opening Exercise

Use the diagram below to complete parts (a)–(c).

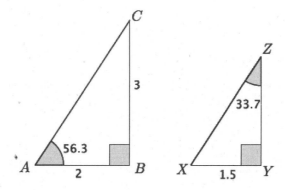

a. Are the triangles shown above similar? Explain.

b. Determine the unknown lengths of the triangles.

c. Explain how you found the lengths in part (a).

Example 1

Recall that an altitude of a triangle is a perpendicular line segment from a vertex to the line determined by the opposite side. In △ ABC to the right, $\overline{BD}$ is the altitude from vertex B to the line containing $\overline{AC}$.

a. How many triangles do you see in the figure?

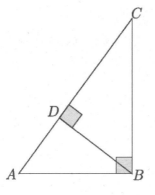

b. Identify the three triangles by name.

We want to consider the altitude of a right triangle from the right angle to the hypotenuse. The altitude of a right triangle splits the triangle into two right triangles, each of which shares a common acute angle with the original triangle. In △ ABC, the altitude $\overline{BD}$ divides the right triangle into two sub-triangles, △ BDC and △ ADB.

c. Is △ ABC ~ △ BDC? Is △ ABC ~ △ ADB? Explain.

d. Is △ABC ~ △DBC? Explain.

e. Since △ABC ~ △BDC and △ABC ~ △ADB, can we conclude that △BDC ~ △ADB? Explain.

f. Identify the altitude drawn in △EFG.

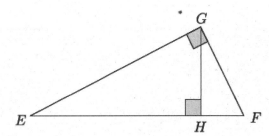

g. As before, the altitude divides the triangle into two sub-triangles, resulting in a total of three triangles including the given triangle. Identify them by name so that the corresponding angles match up.

h. Does the altitude divide △EFG into two similar sub-triangles as the altitude did with △ABC?

The fact that the altitude drawn from the right angle of a right triangle divides the triangle into two similar sub-triangles, which are also similar to the original triangle, allows us to determine the unknown lengths of right triangles.

Example 2

Consider the right triangle △ ABC below.

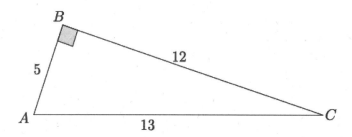

Draw the altitude $\overline{BD}$ from vertex B to the line containing $\overline{AC}$. Label $\overline{AD}$ as x, $\overline{DC}$ as y, and $\overline{BD}$ as z.

Find the values of x, y, and z.

Now we will look at a different strategy for determining the lengths of x, y, and z. The strategy requires that we complete a table of ratios that compares different parts of each triangle.

Make a table of ratios for each triangle that relates the sides listed in the column headers.

	shorter leg: hypotenuse	longer leg: hypotenuse	shorter leg: longer leg
△ ABC			
△ ADB			
△ CDB			

Our work in Example 1 showed us that △ ABC ~ △ ADB ~ △ CDB. Since the triangles are similar, the ratios of their corresponding sides are equal. For example, we can find the length of x by equating the values of shorter leg: hypotenuse ratios of △ ABC and △ ADB.

$$\frac{x}{5} = \frac{5}{13}$$
$$13x = 25$$
$$x = \frac{25}{13} = 1\frac{12}{13}$$

Why can we use these ratios to determine the length of x?

Which ratios can we use to determine the length of y?

Use ratios to determine the length of z.

Since corresponding ratios within similar triangles are equal, we can solve for any unknown side length by equating the values of the corresponding ratios. In the coming lessons, we will learn about more useful ratios for determining unknown side lengths of right triangles.

Lesson 21: Special Relationships Within Right Triangles—Dividing into Two Similar Sub-Triangles

Problem Set

1. Use similar triangles to find the length of the altitudes labeled with variables in each triangle below.

   a.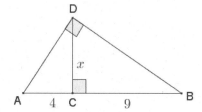

   b.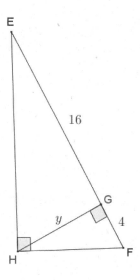

   c.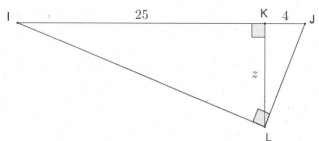

   d. Describe the pattern that you see in your calculations for parts (a) through (c).

2. Given right triangle EFG with altitude $\overline{FH}$ drawn to the hypotenuse, find the lengths of $\overline{EH}$, $\overline{FH}$, and $\overline{GH}$.

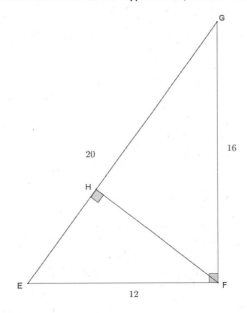

3. Given triangle IMJ with altitude $\overline{JL}$, $JL = 32$, and $IL = 24$, find IJ, JM, LM, and IM.

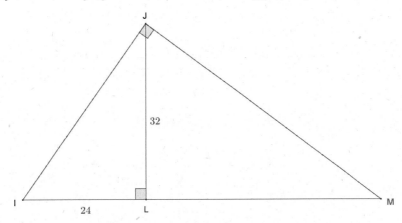

4. Given right triangle RST with altitude $\overline{RU}$ to its hypotenuse, $TU = 1\frac{24}{25}$, and $RU = 6\frac{18}{25}$, find the lengths of the sides of $\triangle RST$.

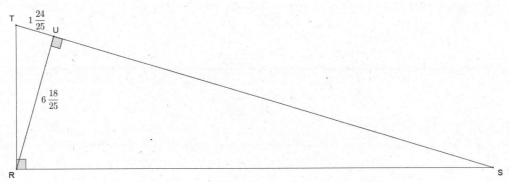

5. Given right triangle ABC with altitude $\overline{CD}$, find AD, BD, AB, and DC.

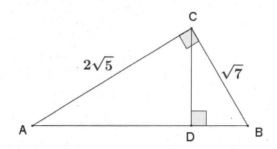

6. Right triangle DEC is inscribed in a circle with radius $AC = 5$. $\overline{DC}$ is a diameter of the circle, $\overline{EF}$ is an altitude of $\triangle DEC$, and $DE = 6$. Find the lengths x and y.

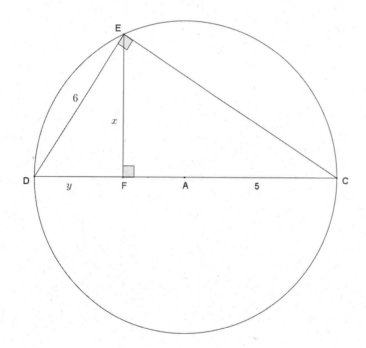

7. In right triangle ABD, $AB = 53$, and altitude $DC = 14$. Find the lengths of $\overline{BC}$ and $\overline{AC}$.

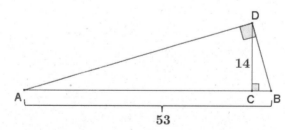

Lesson 22: Multiplying and Dividing Expressions with Radicals

Classwork

Exercises 1–5

Simplify as much as possible.

1. $\sqrt{17^2} =$

2. $\sqrt{5^{10}} =$

3. $\sqrt{4x^4} =$

4. Complete parts (a) through (c).
   a. Compare the value of $\sqrt{36}$ to the value of $\sqrt{9} \times \sqrt{4}$.

b. Make a conjecture about the validity of the following statement: For nonnegative real numbers a and b, $\sqrt{ab} = \sqrt{a} \cdot \sqrt{b}$. Explain.

c. Does your conjecture hold true for $a = -4$ and $b = -9$?

5. Complete parts (a) through (c).

   a. Compare the value of $\sqrt{\dfrac{100}{25}}$ to the value of $\dfrac{\sqrt{100}}{\sqrt{25}}$.

   b. Make a conjecture about the validity of the following statement: For nonnegative real numbers a and b, when $b \neq 0$, $\sqrt{\dfrac{a}{b}} = \dfrac{\sqrt{a}}{\sqrt{b}}$. Explain.

   c. Does your conjecture hold true for $a = -100$ and $b = -25$?

Exercises 6–17

Simplify each expression as much as possible, and rationalize denominators when applicable.

6. $\sqrt{72} =$

7. $\sqrt{\dfrac{17}{25}} =$

8. $\sqrt{32x} =$

9. $\sqrt{\dfrac{1}{3}} =$

10. $\sqrt{54x^2} =$

11. $\dfrac{\sqrt{36}}{\sqrt{18}} =$

12. $\sqrt{\dfrac{4}{x^4}} =$

13. $\dfrac{4x}{\sqrt{64x^2}} =$

14. $\dfrac{5}{\sqrt{x^7}} =$

15. $\sqrt{\dfrac{x^5}{2}} =$

16. $\dfrac{\sqrt{18x}}{3\sqrt{x^5}} =$

17. The captain of a ship recorded the ship's coordinates, then sailed north and then west, and then recorded the new coordinates. The coordinates were used to calculate the distance they traveled, $\sqrt{578}$ km. When the captain asked how far they traveled, the navigator said, "About 24 km." Is the navigator correct? Under what conditions might he need to be more precise in his answer?

Problem Set

Express each number in its simplest radical form.

1. $\sqrt{6} \cdot \sqrt{60} =$

2. $\sqrt{108} =$

3. Pablo found the length of the hypotenuse of a right triangle to be $\sqrt{45}$. Can the length be simplified? Explain.

4. $\sqrt{12x^4} =$

5. Sarahi found the distance between two points on a coordinate plane to be $\sqrt{74}$. Can this answer be simplified? Explain.

6. $\sqrt{16x^3} =$

7. $\dfrac{\sqrt{27}}{\sqrt{3}} =$

8. Nazem and Joffrey are arguing about who got the right answer. Nazem says the answer is $\dfrac{1}{\sqrt{3}}$, and Joffrey says the answer is $\dfrac{\sqrt{3}}{3}$. Show and explain that their answers are equivalent.

9. $\sqrt{\dfrac{5}{8}} =$

10. Determine the area of a square with side length $2\sqrt{7}$ in.

11. Determine the exact area of the shaded region shown below.

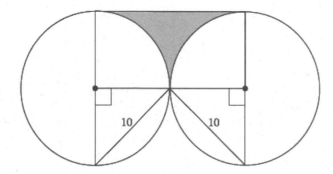

12. Determine the exact area of the shaded region shown to the right.

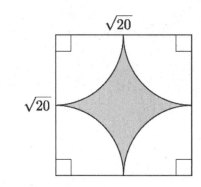

13. Calculate the area of the triangle to the right.

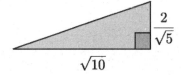

14. $\dfrac{\sqrt{2x^3} \cdot \sqrt{8x}}{\sqrt{x^3}} =$

15. Prove Rule 2 for square roots: $\sqrt{\dfrac{a}{b}} = \dfrac{\sqrt{a}}{\sqrt{b}}$ $(a \geq 0, b > 0)$

    Let p be the nonnegative number so that $p^2 = a$, and let q be the nonnegative number so that $q^2 = b$. Then,

This page intentionally left blank

A STORY OF FUNCTIONS

Lesson 22 M2

GEOMETRY

Perfect Squares of Numbers 1–30

$1^2 = 1$	$16^2 = 256$
$2^2 = 4$	$17^2 = 289$
$3^2 = 9$	$18^2 = 324$
$4^2 = 16$	$19^2 = 361$
$5^2 = 25$	$20^2 = 400$
$6^2 = 36$	$21^2 = 441$
$7^2 = 49$	$22^2 = 484$
$8^2 = 64$	$23^2 = 529$
$9^2 = 81$	$24^2 = 576$
$10^2 = 100$	$25^2 = 625$
$11^2 = 121$	$26^2 = 676$
$12^2 = 144$	$27^2 = 729$
$13^2 = 169$	$28^2 = 784$
$14^2 = 196$	$29^2 = 841$
$15^2 = 225$	$30^2 = 900$

Lesson 22: Multiplying and Dividing Expressions with Radicals

This page intentionally left blank

Lesson 23: Adding and Subtracting Expressions with Radicals

Classwork

Exercises 1–5

Simplify each expression as much as possible.

1. $\sqrt{32} =$

2. $\sqrt{45} =$

3. $\sqrt{300} =$

4. The triangle shown below has a perimeter of $6.5\sqrt{2}$ units. Make a conjecture about how this answer was reached.

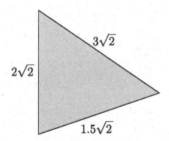

5. The sides of a triangle are $4\sqrt{3}$, $\sqrt{12}$, and $\sqrt{75}$. Make a conjecture about how to determine the perimeter of this triangle.

Exercise 6

6. Circle the expressions that can be simplified using the distributive property. Be prepared to explain your choices.

$8.3\sqrt{2} + 7.9\sqrt{2}$
$\sqrt{13} - \sqrt{6}$
$-15\sqrt{5} + \sqrt{45}$
$11\sqrt{7} - 6\sqrt{7} + 3\sqrt{2}$
$19\sqrt{2} + 2\sqrt{8}$
$4 + \sqrt{11}$
$\sqrt{7} + 2\sqrt{10}$
$\sqrt{12} - \sqrt{75}$
$\sqrt{32} + \sqrt{2}$
$6\sqrt{13} + \sqrt{26}$

Example 1

Explain how the expression $8.3\sqrt{2} + 7.9\sqrt{2}$ can be simplified using the distributive property.

Explain how the expression $11\sqrt{7} - 6\sqrt{7} + 3\sqrt{2}$ can be simplified using the distributive property.

Lesson 23: Adding and Subtracting Expressions with Radicals

Example 2

Explain how the expression $19\sqrt{2} + 2\sqrt{8}$ can be simplified using the distributive property.

Example 3

Can the expression $\sqrt{7} + 2\sqrt{10}$ be simplified using the distributive property?

To determine if an expression can be simplified, you must first simplify each of the terms within the expression. Then, apply the distributive property, or other properties as needed, to simplify the expression.

Lesson 23

Problem Set

Express each answer in simplified radical form.

1. $18\sqrt{5} - 12\sqrt{5} =$

2. $\sqrt{24} + 4\sqrt{54} =$

3. $2\sqrt{7} + 4\sqrt{63} =$

4. What is the perimeter of the triangle shown below?

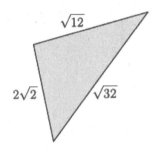

5. Determine the area and perimeter of the triangle shown. Simplify as much as possible.

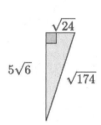

6. Determine the area and perimeter of the rectangle shown. Simplify as much as possible.

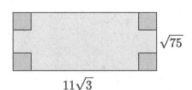

7. Determine the area and perimeter of the triangle shown. Simplify as much as possible.

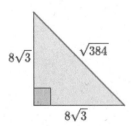

Lesson 23: Adding and Subtracting Expressions with Radicals

8. Determine the area and perimeter of the triangle shown. Simplify as much as possible.

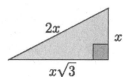

9. The area of the rectangle shown in the diagram below is 160 square units. Determine the area and perimeter of the shaded triangle. Write your answers in simplest radical form, and then approximate to the nearest tenth.

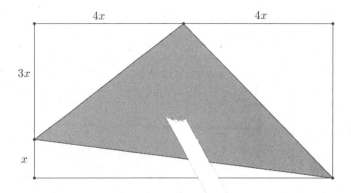

10. Parallelogram $ABCD$ has an area of $9\sqrt{3}$ square units. $DC = 3\sqrt{3}$, and G and H are midpoints of $\overline{DE}$ and $\overline{CE}$, respectively. Find the area of the shaded region. Write your answer in simplest radical form.

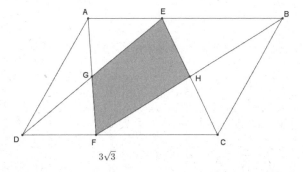

Lesson 23: Adding and Subtracting Expressions with Radicals

This page intentionally left blank

Lesson 24: Prove the Pythagorean Theorem Using Similarity

Classwork

Exercises 1–3

1. Find the length of the hypotenuse of a right triangle whose legs have lengths 50 and 100.

2. Can you think of a simpler method for finding the length of the hypotenuse in Exercise 1? Explain.

3. Find the length of the hypotenuse of a right triangle whose legs have lengths 75 and 225.

Exploratory Challenge/Exercises 4–5

4. An equilateral triangle has sides of length 2 and angle measures of 60°, as shown below. The altitude from one vertex to the opposite side divides the triangle into two right triangles.

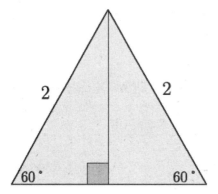

a. Are those triangles congruent? Explain.

b. What is the length of the shorter leg of each of the right triangles? Explain.

c. Use the Pythagorean theorem to determine the length of the altitude.

d. Write the ratio that represents shorter leg: hypotenuse.

e. Write the ratio that represents longer leg: hypotenuse.

f. Write the ratio that represents shorter leg: longer leg.

g. By the AA criterion, any triangles with measures 30–60–90 will be similar to this triangle. If a 30–60–90 triangle has a hypotenuse of length 16, what are the lengths of the legs?

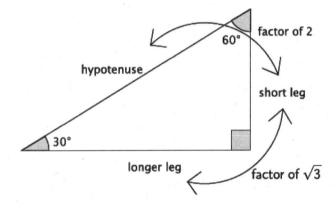

5. An isosceles right triangle has leg lengths of 1, as shown.

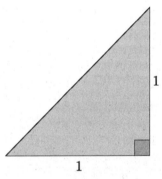

a. What are the measures of the other two angles? Explain.

b. Use the Pythagorean theorem to determine the length of the hypotenuse of the right triangle.

c. Is it necessary to write all three ratios: shorter leg: hypotenuse, longer leg: hypotenuse, and shorter leg: longer leg? Explain.

d. Write the ratio that represents leg: hypotenuse.

e. By the AA criterion, any triangles with measures 45–45–90 will be similar to this triangle. If a 45–45–90 triangle has a hypotenuse of length 20, what are the lengths of the legs?

Problem Set

1. In each row of the table below are the lengths of the legs and hypotenuses of different right triangles. Find the missing side lengths in each row, in simplest radical form.

Leg$_1$	Leg$_2$	Hypotenuse
15		25
15	36	
3		7
100	200	

2. Claude sailed his boat due south for 38 miles and then due west for 25 miles. Approximately how far is Claude from where he began?

3. Find the lengths of the legs in the triangle given the hypotenuse with length 100.

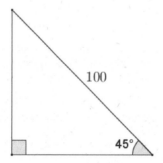

4. Find the length of the hypotenuse in the right triangle given that the legs have lengths of 100.

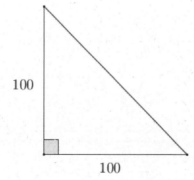

5. Each row in the table below shows the side lengths of a different 30–60–90 right triangle. Complete the table with the missing side lengths in simplest radical form. Use the relationships of the values in the first three rows to complete the last row. How could the expressions in the last row be used?

Shorter Leg	Longer Leg	Hypotenuse
25		50
15		
	3	$2\sqrt{3}$
x		

6. In right triangle ABC with $\angle C$ a right angle, an altitude of length h is dropped to side $\overline{AB}$ that splits the side $\overline{AB}$ into segments of length x and y. Use the Pythagorean theorem to show $h^2 = xy$.

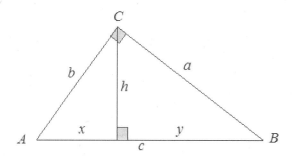

7. In triangle ABC, the altitude from $\angle C$ splits side $\overline{AB}$ into two segments of lengths x and y. If h denotes the length of the altitude and $h^2 = xy$, use the Pythagorean theorem and its converse to show that triangle ABC is a right triangle with $\angle C$ a right angle.

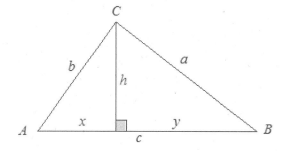

Lesson 24: Prove the Pythagorean Theorem Using Similarity

Lesson 25: Incredibly Useful Ratios

Classwork

Exercises 1–3

Use the right triangle △ ABC to answer Exercises 1–3.

1. Name the side of the triangle opposite ∠A.

2. Name the side of the triangle opposite ∠B.

3. Name the side of the triangle opposite ∠C.

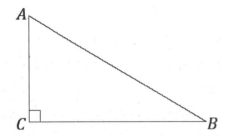

Exercises 4–6

For each exercise, label the appropriate sides as *adjacent, opposite,* and *hypotenuse,* with respect to the marked acute angle.

4.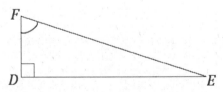

5.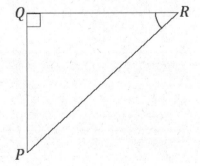

6.

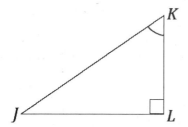

Exploratory Challenge

Note: Angle measures are approximations.

For each triangle in your set, determine missing angle measurements and side lengths. Side lengths should be measured to one decimal place. Make sure that each of the $\frac{\text{adj}}{\text{hyp}}$ and $\frac{\text{opp}}{\text{hyp}}$ ratios are set up and missing values are calculated and rounded appropriately.

	Group 1				
	Triangle	Angle Measures	Length Measures	$\frac{\text{opp}}{\text{hyp}}$	$\frac{\text{adj}}{\text{hyp}}$
1.	△ ABC			$\frac{12}{13} \approx 0.92$	$\frac{5}{13} \approx 0.38$
2.	△ DEF	$m\angle D \approx 53°$	$DE = 3$ cm $EF = 4$ cm $DF = 5$ cm		
3.	△ GHI	$m\angle I \approx 41°$	$GH = 5.3$ cm	$\frac{5.3}{} \approx 0.66$	$= 0.75$
4.	△ JKL		$KL = 6.93$ cm $JL = 8$ cm	$\frac{}{8} =$	$\frac{}{8} \approx 0.87$
5.	△ MNO			$\frac{4}{8.5} \approx 0.47$	$\frac{7.5}{8.5} \approx 0.88$

Lesson 25: Incredibly Useful Ratios

Group 2

	Triangle	Angle Measures	Length Measures	$\dfrac{\text{opp}}{\text{hyp}}$	$\dfrac{\text{adj}}{\text{hyp}}$
1.	$\triangle A'B'C'$			$\dfrac{6}{6.5} \approx 0.92$	$\dfrac{2.5}{6.5} \approx 0.38$
2.	$\triangle D'E'F'$	$m\angle D' \approx 53°$	$D'E' = 6$ cm $E'F' = 8$ cm $D'F' = 10$ cm		
3.	$\triangle G'H'I'$	$m\angle I' \approx 41°$	$G'H' = 7.9$ cm	$\dfrac{7.9}{\_\_} \approx 0.66$	$\_\_ = 0.75$
4.	$\triangle J'K'L'$		$K'L' = 10.4$ cm $J'L' = 12$ cm	$\dfrac{__}{12} = $	$\dfrac{\_\_}{12} \approx 0.87$
5.	$\triangle M'N'O'$			$\dfrac{8}{17} \approx 0.47$	$\dfrac{15}{17} \approx 0.88$

With a partner, discuss what you can conclude about each pair of triangles between the two sets.

Exercises 7–10

For each question, round the unknown lengths appropriately. Refer back to your completed chart from the Exploratory Challenge; each indicated acute angle is the same approximated acute angle measure as in the chart. Set up and label the appropriate length ratios, using the terms opp, adj, and hyp in the setup of each ratio.

7.

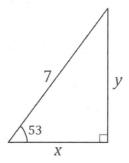

8.

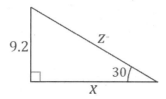

9.

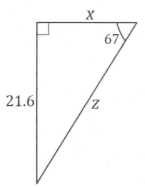

10. From a point 120 m away from a building, Serena measures the angle between the ground and the top of a building and finds it measures 41°.

    What is the height of the building? Round to the nearest meter.

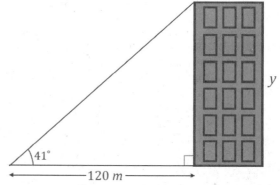

Lesson 25: Incredibly Useful Ratios

Problem Set

The table below contains the values of the ratios $\frac{\text{opp}}{\text{hyp}}$ and $\frac{\text{adj}}{\text{hyp}}$ for a variety of right triangles based on a given acute angle, θ, from each triangle. Use the table and the diagram of the right triangle below to complete each problem.

θ (degrees)	0	10	20	30	40	45	50	60	70	80	90
$\frac{\text{opp}}{\text{hyp}}$	0	0.1736	0.3420	$\frac{1}{2} = 0.5$	0.6428	0.7071	0.7660	0.8660	0.9397	0.9848	1
$\frac{\text{adj}}{\text{hyp}}$	1	0.9848	0.9397	0.8660	0.7660	0.7071	0.6428	$\frac{1}{2} = 0.5$	0.3420	0.1736	0

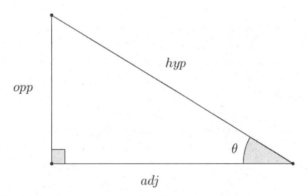

NOT DRAWN TO SCALE

For each problem, approximate the unknown lengths to one decimal place. Write the appropriate length ratios using the terms opp, adj, and hyp in the setup of each ratio.

1. Find the approximate length of the leg opposite the 80° angle.

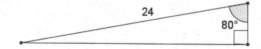

2. Find the approximate length of the hypotenuse.

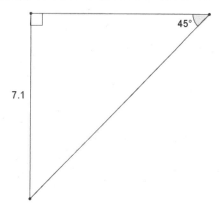

3. Find the approximate length of the hypotenuse.

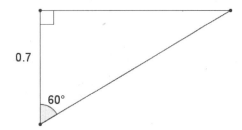

4. Find the approximate length of the leg adjacent to the 40° angle.

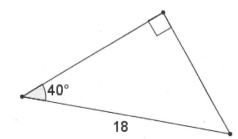

5. Find the length of both legs of the right triangle below. Indicate which leg is adjacent and which is opposite the given angle of 30°.

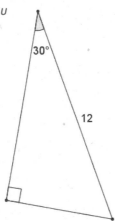

6. Three city streets form a right triangle. Main Street and State Street are perpendicular. Laura Street and State Street intersect at a 50° angle. The distance along Laura Street to Main Street is 0.8 mile. If Laura Street is closed between Main Street and State Street for a festival, approximately how far (to the nearest tenth) will someone have to travel to get around the festival if they take only Main Street and State Street?

7. A cable anchors a utility pole to the ground as shown in the picture. The cable forms an angle of 70° with the ground. The distance from the base of the utility pole to the anchor point on the ground is 3.8 meters. Approximately how long is the support cable?

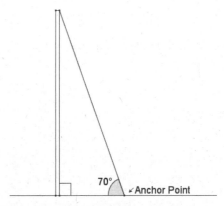

8. Indy says that the ratio of $\frac{opp}{adj}$ for an angle of 0° has a value of 0 because the opposite side of the triangle has a length of 0. What does she mean?

Lesson 25: Incredibly Useful Ratios

This page intentionally left blank

A STORY OF FUNCTIONS

Lesson 25 M2

GEOMETRY

Group 1

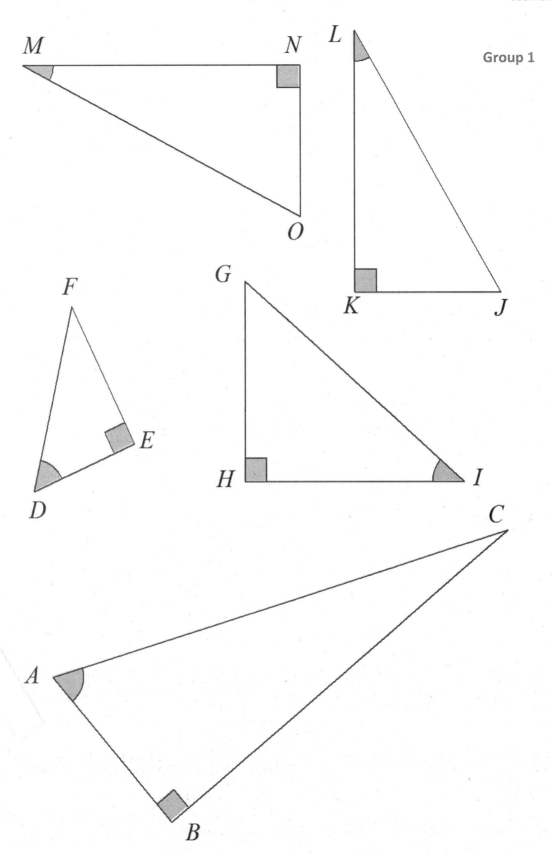

Lesson 25: Incredibly Useful Ratios

This page intentionally left blank

A STORY OF FUNCTIONS   **Lesson 25**   **M2**

GEOMETRY

Group 2

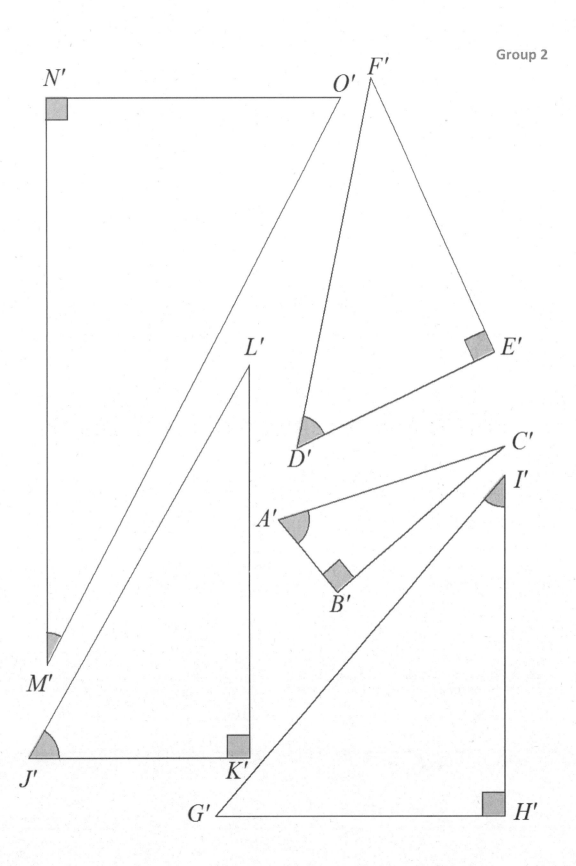

Lesson 25:   Incredibly Useful Ratios

This page intentionally left blank

Identifying Sides of a Right Triangle with Respect to a Given Acute Angle

Poster

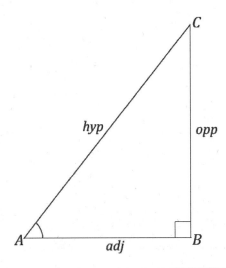 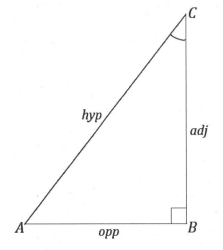

• With respect to $\angle A$, the opposite side, opp, is side $\overline{BC}$. • With respect to $\angle A$, the adjacent side, adj, is side $\overline{AB}$. • The hypotenuse, hyp, is side $\overline{AC}$ and is always opposite from the 90° angle.	• With respect to $\angle C$, the opposite side, opp, is side $\overline{AB}$. • With respect to $\angle C$, the adjacent side, adj, is side $\overline{BC}$. • The hypotenuse, hyp, is side $\overline{AC}$ and is always opposite from the 90° angle.

This page intentionally left blank

Lesson 26: Definition of Sine, Cosine, and Tangent

Classwork

Exercises 1–3

1. Identify the $\frac{\text{opp}}{\text{hyp}}$ ratios for $\angle A$ and $\angle B$.

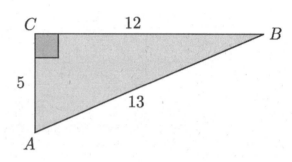

2. Identify the $\frac{\text{adj}}{\text{hyp}}$ ratios for $\angle A$ and $\angle B$.

3. Describe the relationship between the ratios for $\angle A$ and $\angle B$.

Exercises 4–9

4. In △ PQR, $m\angle P = 53.2°$ and $m\angle Q = 36.8°$. Complete the following table.

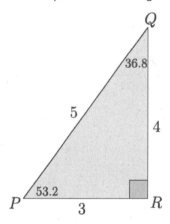

Measure of Angle	Sine $\left(\dfrac{\text{opp}}{\text{hyp}}\right)$	Cosine $\left(\dfrac{\text{adj}}{\text{hyp}}\right)$	Tangent $\left(\dfrac{\text{opp}}{\text{adj}}\right)$
53.2			
36.8			

5. In the triangle below, $m\angle A = 33.7°$ and $m\angle B = 56.3°$. Complete the following table.

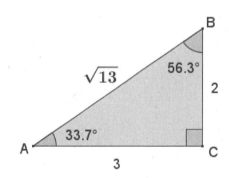

Measure of Angle	Sine	Cosine	Tangent
33.7			
56.3			

Lesson 26: Definition of Sine, Cosine, and Tangent

6. In the triangle below, let e be the measure of $\angle E$ and d be the measure of $\angle D$. Complete the following table.

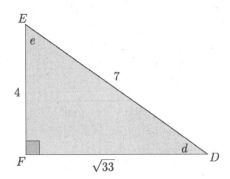

Measure of Angle	Sine	Cosine	Tangent
d			
e			

7. In the triangle below, let x be the measure of $\angle X$ and y be the measure of $\angle Y$. Complete the following table.

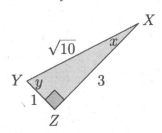

Measure of Angle	Sine	Cosine	Tangent
x			
y			

Lesson 26: Definition of Sine, Cosine, and Tangent

8. Tamer did not finish completing the table below for a diagram similar to the previous problems that the teacher had on the board where p was the measure of $\angle P$ and q was the measure of $\angle Q$. Use any patterns you notice from Exercises 1–4 to complete the table for Tamer.

Measure of Angle	Sine	Cosine	Tangent
p	$\sin p = \dfrac{11}{\sqrt{157}}$	$\cos p = \dfrac{6}{\sqrt{157}}$	$\tan p = \dfrac{11}{6}$
q			

9. Explain how you were able to determine the sine, cosine, and tangent of $\angle Q$ in Exercise 8.

Problem Set

1. Given the triangle in the diagram, complete the following table.

Angle Measure	sin	cos	tan
α			
β			

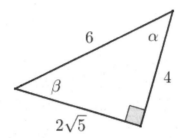

2. Given the table of values below (not in simplest radical form), label the sides and angles in the right triangle.

Angle Measure	sin	cos	tan
α	$\dfrac{4}{2\sqrt{10}}$	$\dfrac{2\sqrt{6}}{2\sqrt{10}}$	$\dfrac{4}{2\sqrt{6}}$
β	$\dfrac{2\sqrt{6}}{2\sqrt{10}}$	$\dfrac{4}{2\sqrt{10}}$	$\dfrac{2\sqrt{6}}{4}$

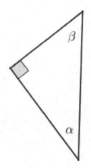

Lesson 26: Definition of Sine, Cosine, and Tangent

3. Given sin α and sin β, complete the missing values in the table. You may draw a diagram to help you.

Angle Measure	sin	cos	tan
α	$\dfrac{\sqrt{2}}{3\sqrt{3}}$	$\dfrac{5}{3\sqrt{3}}$	
β			

4. Given the triangle shown to the right, fill in the missing values in the table.

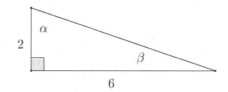

Angle Measure	sin	cos	tan
α			
β			

5. Jules thinks that if α and β are two different acute angle measures, then sin α ≠ sin β. Do you agree or disagree? Explain.

6. Given the triangle in the diagram, complete the following table.

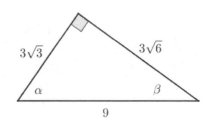

Angle Measure	sin	cos	tan
α			
β			

Lesson 26: Definition of Sine, Cosine, and Tangent

Rewrite the values from the table in simplest terms.

Angle Measure	sin	cos	tan
α			
β			

Draw and label the sides and angles of a right triangle using the values of the ratios sin and cos. How is the new triangle related to the original triangle?

7. Given $\tan \alpha$ and $\cos \beta$, in simplest terms, find the missing side lengths of the right triangle if one leg of the triangle has a length of 4. Draw and label the sides and angles of the right triangle.

Angle Measure	$\sin \theta$	$\cos \theta$	$\tan \theta$
α			
β			

8. Eric wants to hang a rope bridge over a small ravine so that it is easier to cross. To hang the bridge, he needs to know how much rope is needed to span the distance between two trees that are directly across from each other on either side of the ravine. Help Eric devise a plan using sine, cosine, and tangent to determine the approximate distance from tree A to tree B without having to cross the ravine.

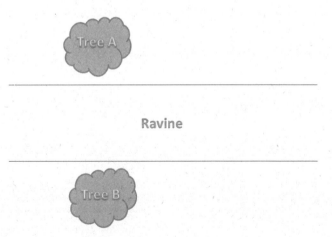

Lesson 26: Definition of Sine, Cosine, and Tangent

9. A fisherman is at point F on the open sea and has three favorite fishing locations. The locations are indicated by points A, B, and C. The fisherman plans to sail from F to A, then to B, then to C, and then back to F. If the fisherman is 14 miles from $\overline{AC}$, find the total distance that he will sail.

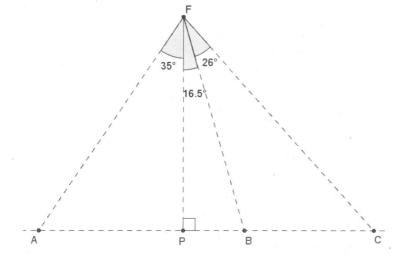

Lesson 27: Sine and Cosine of Complementary and Special Angles

Classwork

Example 1

If α and β are the measurements of complementary angles, then we are going to show that $\sin \alpha = \cos \beta$.

In right triangle ABC, the measurement of acute angle $\angle A$ is denoted by α, and the measurement of acute angle $\angle B$ is denoted by β.

Determine the following values in the table:

$\sin \alpha$	$\sin \beta$	$\cos \alpha$	$\cos \beta$

What can you conclude from the results?

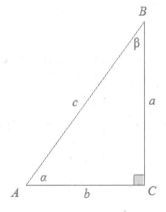

Exercises 1–3

1. Consider the right triangle ABC so that $\angle C$ is a right angle, and the degree measures of $\angle A$ and $\angle B$ are α and β, respectively.

    a. Find $\alpha + \beta$.

    b. Use trigonometric ratios to describe $\dfrac{BC}{AB}$ two different ways.

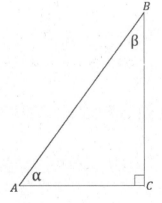

c. Use trigonometric ratios to describe $\dfrac{AC}{AB}$ two different ways.

d. What can you conclude about $\sin \alpha$ and $\cos \beta$?

e. What can you conclude about $\cos \alpha$ and $\sin \beta$?

2. Find values for θ that make each statement true.
   a. $\sin \theta = \cos(25)$

   b. $\sin 80 = \cos \theta$

   c. $\sin \theta = \cos(\theta + 10)$

   d. $\sin(\theta - 45) = \cos(\theta)$

3. For what angle measurement must sine and cosine have the same value? Explain how you know.

Example 2

What is happening to a and b as θ changes? What happens to $\sin\theta$ and $\cos\theta$?

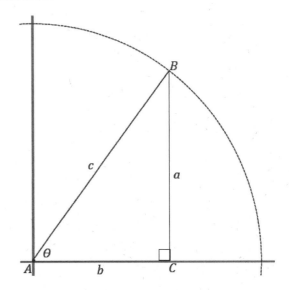

Example 3

There are certain special angles where it is possible to give the exact value of sine and cosine. These are the angles that measure 0°, 30°, 45°, 60°, and 90°; these angle measures are frequently seen.

You should memorize the sine and cosine of these angles with quick recall just as you did your arithmetic facts.

a. Learn the following sine and cosine values of the key angle measurements.

θ	0°	30°	45°	60°	90°
Sine	0	$\dfrac{1}{2}$	$\dfrac{\sqrt{2}}{2}$	$\dfrac{\sqrt{3}}{2}$	1
Cosine	1	$\dfrac{\sqrt{3}}{2}$	$\dfrac{\sqrt{2}}{2}$	$\dfrac{1}{2}$	0

We focus on an easy way to remember the entries in the table. What do you notice about the table values?

This is easily explained because the pairs (0,90), (30,60), and (45,45) are the measures of complementary angles. So, for instance, $\sin 30 = \cos 60$.

The sequence $0, \dfrac{1}{2}, \dfrac{\sqrt{2}}{2}, \dfrac{\sqrt{3}}{2}, 1$ may be easier to remember as the sequence $\dfrac{\sqrt{0}}{2}, \dfrac{\sqrt{1}}{2}, \dfrac{\sqrt{2}}{2}, \dfrac{\sqrt{3}}{2}, \dfrac{\sqrt{4}}{2}$.

Lesson 27: Sine and Cosine of Complementary and Special Angles

b. △ ABC is equilateral, with side length 2; D is the midpoint of side $\overline{AC}$. Label all side lengths and angle measurements for △ ABD. Use your figure to determine the sine and cosine of 30 and 60.

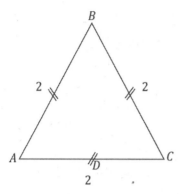

c. Draw an isosceles right triangle with legs of length 1. What are the measures of the acute angles of the triangle? What is the length of the hypotenuse? Use your triangle to determine sine and cosine of the acute angles.

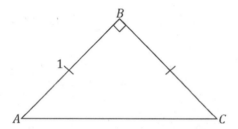

Parts (b) and (c) demonstrate how the sine and cosine values of the mentioned special angles can be found. These triangles are common to trigonometry; we refer to the triangle in part (b) as a 30–60–90 triangle and the triangle in part (c) as a 45–45–90 triangle.

30–60–90 Triangle, side length ratio $1:2:\sqrt{3}$	45–45–90 Triangle, side length ratio $1:1:\sqrt{2}$
$2:4:2\sqrt{3}$	$2:2:2\sqrt{2}$
$3:6:3\sqrt{3}$	$3:3:3\sqrt{2}$
$4:8:4\sqrt{3}$	$4:4:4\sqrt{2}$
$x:2x:x\sqrt{3}$	$x:x:x\sqrt{2}$

Exercises 4–5

4. Find the missing side lengths in the triangle.

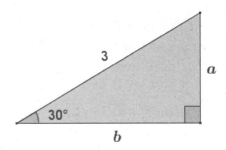

5. Find the missing side lengths in the triangle.

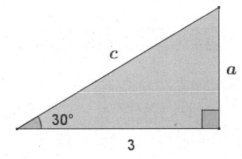

Problem Set

1. Find the value of θ that makes each statement true.
   a. $\sin\theta = \cos(\theta + 38)$
   b. $\cos\theta = \sin(\theta - 30)$
   c. $\sin\theta = \cos(3\theta + 20)$
   d. $\sin\left(\dfrac{\theta}{3} + 10\right) = \cos\theta$

2.
   a. Make a prediction about how the sum $\sin 30 + \cos 60$ will relate to the sum $\sin 60 + \cos 30$.
   b. Use the sine and cosine values of special angles to find the sum: $\sin 30 + \cos 60$.
   c. Find the sum: $\sin 60 + \cos 30$.
   d. Was your prediction a valid prediction? Explain why or why not.

3. Langdon thinks that the sum $\sin 30 + \sin 30$ is equal to $\sin 60$. Do you agree with Langdon? Explain what this means about the sum of the sines of angles.

4. A square has side lengths of $7\sqrt{2}$. Use sine or cosine to find the length of the diagonal of the square. Confirm your answer using the Pythagorean theorem.

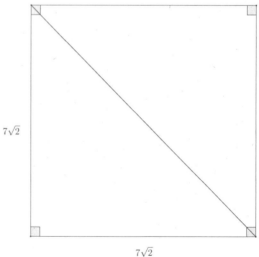

5. Given an equilateral triangle with sides of length 9, find the length of the altitude. Confirm your answer using the Pythagorean theorem.

Lesson 28: Solving Problems Using Sine and Cosine

Classwork

Exercises 1–4

1.
   a. The bus drops you off at the corner of H Street and 1st Street, approximately 300 ft. from school. You plan to walk to your friend Janneth's house after school to work on a project. Approximately how many feet will you have to walk from school to Janneth's house? Round your answer to the nearest foot. (Hint: Use the ratios you developed in Lesson 25.)

   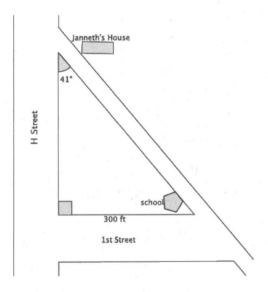

   b. In real life, it is unlikely that you would calculate the distance between school and Janneth's house in this manner. Describe a similar situation in which you might actually want to determine the distance between two points using a trigonometric ratio.

2. Use a calculator to find the sine and cosine of θ. Give your answer rounded to the ten-thousandth place.

θ	0	10	20	30	40	50	60	70	80	90
$\sin \theta$										
$\cos \theta$										

3. What do you notice about the numbers in the row $\sin \theta$ compared with the numbers in the row $\cos \theta$?

4. Provide an explanation for what you noticed in Exercise 2.

Example 1

Find the values of a and b.

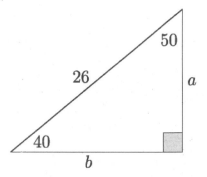

Exercise 5

5. A shipmate set a boat to sail exactly 27° NE from the dock. After traveling 120 miles, the shipmate realized he had misunderstood the instructions from the captain; he was supposed to set sail going directly east!

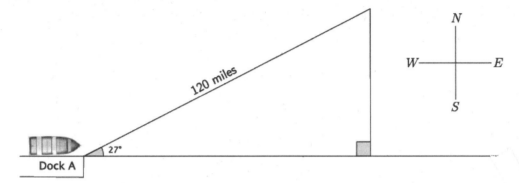

a. How many miles will the shipmate have to travel directly south before he is directly east of the dock? Round your answer to the nearest mile.

b. How many extra miles does the shipmate travel by going the wrong direction compared to going directly east? Round your answer to the nearest mile.

Lesson 28: Solving Problems Using Sine and Cosine

Example 2

Johanna borrowed some tools from a friend so that she could precisely, but not exactly, measure the corner space in her backyard to plant some vegetables. She wants to build a fence to prevent her dog from digging up the seeds that she plants. Johanna returned the tools to her friend before making the most important measurement: the one that would give the length of the fence!

Johanna decided that she could just use the Pythagorean theorem to find the length of the fence she would need. Is the Pythagorean theorem applicable in this situation? Explain.

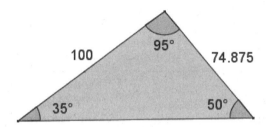

Exercise 6

6. The measurements of the triangle shown below are rounded to the nearest hundredth. Calculate the missing side length to the nearest hundredth.

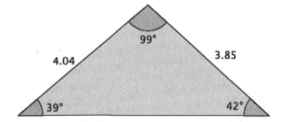

Problem Set

1. Given right triangle GHI, with right angle at H, $GH = 12.2$, and $m\angle G = 28°$, find the measures of the remaining sides and angle to the nearest tenth.

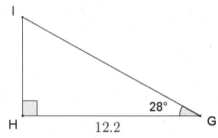

2. The Occupational Safety and Health Administration (OSHA) provides standards for safety at the workplace. A ladder is leaned against a vertical wall according to OSHA standards and forms an angle of approximately 75° with the floor.

   a. If the ladder is 25 ft. long, what is the distance from the base of the ladder to the base of the wall?
   b. How high on the wall does the ladder make contact?
   c. Describe how to safely set a ladder according to OSHA standards without using a protractor.

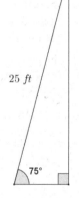

3. A regular pentagon with side lengths of 14 cm is inscribed in a circle. What is the radius of the circle?

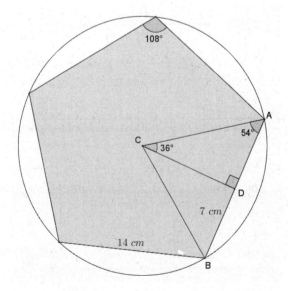

Lesson 28: Solving Problems Using Sine and Cosine

4. The circular frame of a Ferris wheel is suspended so that it sits 4 ft. above the ground and has a radius of 30 ft. A segment joins center C to point S on the circle. If $\overline{CS}$ makes an angle of 48° with the horizon, what is the distance of point S to the ground?

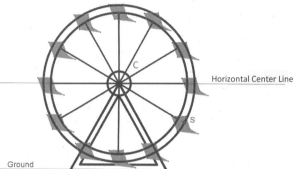

5. Tim is a contractor who is designing a wheelchair ramp for handicapped access to a business. According to the Americans with Disabilities Act (ADA), the maximum slope allowed for a public wheelchair ramp forms an angle of approximately 4.76° to level ground. The length of a ramp's surface cannot exceed 30 ft. without including a flat 5 ft. × 5 ft. platform (minimum dimensions) on which a person can rest, and such a platform must be included at the bottom and top of any ramp.

   Tim designs a ramp that forms an angle of 4° to the level ground to reach the entrance of the building. The entrance of the building is 2 ft. 9 in. above the ground. Let x and y as shown in Tim's initial design below be the indicated distances in feet.

   a. Assuming that the ground in front of the building's entrance is flat, use Tim's measurements and the ADA requirements to complete and/or revise his wheelchair ramp design.

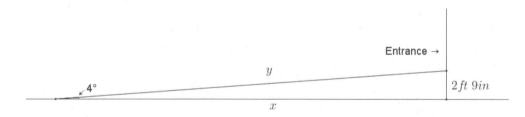

   (For more information, see section 405 of the 2010 ADA Standards for Accessible Design at the following link: http://www.ada.gov/regs2010/2010ADAStandards/2010ADAstandards.htm#pgfld-1006877.)

   b. What is the total distance from the start of the ramp to the entrance of the building in your design?

6. Tim is designing a roof truss in the shape of an isosceles triangle. The design shows the base angles of the truss to have measures of 18.5°. If the horizontal base of the roof truss is 36 ft. across, what is the height of the truss?

Lesson 29: Applying Tangents

Classwork

Opening Exercise

a. Use a calculator to find the tangent of θ. Enter the values, correct to four decimal places, in the last row of the table.

θ	0	10	20	30	40	50	60	70	80	90
$\sin\theta$	0	0.1736	0.3420	0.5	0.6428	0.7660	0.8660	0.9397	0.9848	1
$\cos\theta$	1	0.9848	0.9397	0.8660	0.7660	0.6428	0.5	0.3420	0.1736	0
$\dfrac{\sin\theta}{\cos\theta}$										
$\tan\theta$										

b. The table from Lesson 29 is provided here for you. In the row labeled $\dfrac{\sin\theta}{\cos\theta}$, divide the sine values by the cosine values. What do you notice?

Lesson 29: Applying Tangents

S.219

Example 1

Scott, whose eye level is 1.5 m above the ground, stands 30 m from a tree. The angle of elevation of a bird at the top of the tree is 36°. How far above ground is the bird?

Example 2

From an angle of depression of 40°, John watches his friend approach his building while standing on the rooftop. The rooftop is 16 m from the ground, and John's eye level is at about 1.8 m from the rooftop. What is the distance between John's friend and the building?

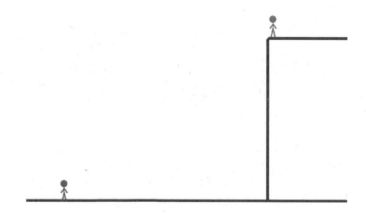

Exercise 1

Standing on the gallery of a lighthouse (the deck at the top of a lighthouse), a person spots a ship at an angle of depression of 20°. The lighthouse is 28 m tall and sits on a cliff 45 m tall as measured from sea level. What is the horizontal distance between the lighthouse and the ship? Sketch a diagram to support your answer.

Exercise 2

A line on the coordinate plane makes an angle of depression of 36°. Find the slope of the line correct to four decimal places.

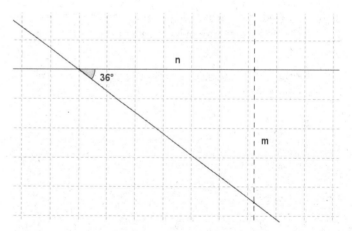

Problem Set

1. A line in the coordinate plane has an angle of elevation of 53°. Find the slope of the line correct to four decimal places.

2. A line in the coordinate plane has an angle of depression of 25°. Find the slope of the line correct to four decimal places.

3. In Problems 1 and 2, why do the lengths of the legs of the right triangles formed not affect the slope of the line?

4. Given the angles of depression below, determine the slope of the line with the indicated angle correct to four decimal places.
   a. 35° angle of depression
   b. 49° angle of depression
   c. 80° angle of depression
   d. 87° angle of depression
   e. 89° angle of depression
   f. 89.9° angle of depression
   g. What appears to be happening to the slopes (and tangent values) as the angles of depression get closer to 90°?
   h. Find the slopes of angles of depression that are even closer to 90° than 89.9°. Can the value of the tangent of 90° be defined? Why or why not?

5. For the indicated angle, express the quotient in terms of sine, cosine, or tangent. Then, write the quotient in simplest terms.
   a. $\dfrac{4}{2\sqrt{13}}$; α
   b. $\dfrac{6}{4}$; α
   c. $\dfrac{4}{2\sqrt{13}}$; β
   d. $\dfrac{4}{6}$; β

6. The pitch of a roof on a home is expressed as a ratio of vertical rise: horizontal run where the run has a length of 12 units. If a given roof design includes an angle of elevation of 22.5° and the roof spans 36 ft. as shown in the diagram, determine the pitch of the roof. Then, determine the distance along one of the two sloped surfaces of the roof.

Lesson 29: Applying Tangents

7. An anchor cable supports a vertical utility pole forming a 51° angle with the ground. The cable is attached to the top of the pole. If the distance from the base of the pole to the base of the cable is 5 meters, how tall is the pole?

8. A winch is a tool that rotates a cylinder, around which a cable is wound. When the winch rotates in one direction, it draws the cable in. Joey is using a winch and a pulley (as shown in the diagram) to raise a heavy box off the floor and onto a cart. The box is 2 ft. tall, and the winch is 14 ft. horizontally from where cable drops down vertically from the pulley. The angle of elevation to the pulley is 42°. What is the approximate length of cable required to connect the winch and the box?

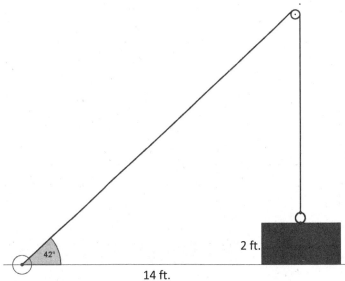

Lesson 29: Applying Tangents

Lesson 30: Trigonometry and the Pythagorean Theorem

Classwork

Exercises 1–2

1. In a right triangle with acute angle of measure θ, $\sin\theta = \frac{1}{2}$. What is the value of $\cos\theta$? Draw a diagram as part of your response.

2. In a right triangle with acute angle of measure θ, $\sin\theta = \frac{7}{9}$. What is the value of $\tan\theta$? Draw a diagram as part of your response.

Example 1

a. What common right triangle was probably modeled in the construction of the triangle in Figure 2? Use $\sin 53° \approx 0.8$.

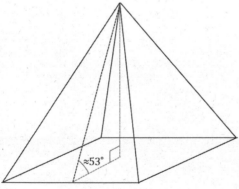

Figure 1

b. The actual angle between the base and lateral faces of the pyramid is actually closer to 52°. Considering the age of the pyramid, what could account for the difference between the angle measure in part (a) and the actual measure?

c. Why do you think the architects chose to use a 3–4–5 as a model for the triangle?

Example 2

Show why $\tan \theta = \dfrac{\sin \theta}{\cos \theta}$

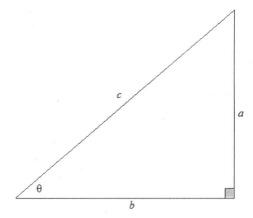

Exercises 3–4

3. In a right triangle with acute angle of measure θ, $\sin\theta = \frac{1}{2}$, use the Pythagorean identity to determine the value of $\cos\theta$.

4. Given a right triangle with acute angle of measure θ, $\sin\theta = \frac{7}{9}$, use the Pythagorean identity to determine the value of $\tan\theta$.

Problem Set

1. If $\cos\theta = \frac{4}{5}$, find $\sin\theta$ and $\tan\theta$.

2. If $\sin\theta = \frac{44}{125}$, find $\cos\theta$ and $\tan\theta$.

3. If $\tan\theta = 5$, find $\sin\theta$ and $\cos\theta$.

4. If $\sin\theta = \frac{\sqrt{5}}{5}$, find $\cos\theta$ and $\tan\theta$.

5. Find the missing side lengths of the following triangle using sine, cosine, and/or tangent. Round your answer to four decimal places.

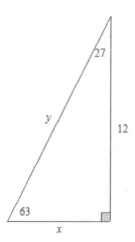

6. A surveying crew has two points A and B marked along a roadside at a distance of 400 yd. A third point C is marked at the back corner of a property along a perpendicular to the road at B. A straight path joining C to A forms a 28° angle with the road. Find the distance from the road to point C at the back of the property and the distance from A to C using sine, cosine, and/or tangent. Round your answer to three decimal places.

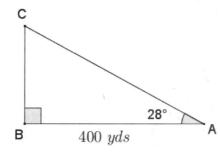

Lesson 30: Trigonometry and the Pythagorean Theorem

7. The right triangle shown is taken from a slice of a right rectangular pyramid with a square base.
   a. Find the height of the pyramid (to the nearest tenth).
   b. Find the lengths of the sides of the base of the pyramid (to the nearest tenth).
   c. Find the lateral surface area of the right rectangular pyramid.

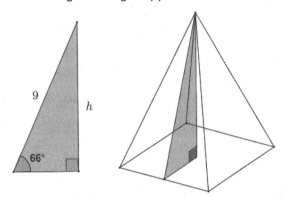

8. A machinist is fabricating a wedge in the shape of a right triangular prism. One acute angle of the right triangular base is 33°, and the opposite side is 6.5 cm. Find the length of the edges labeled l and m using sine, cosine, and/or tangent. Round your answer to the nearest thousandth of a centimeter.

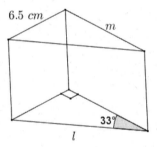

9. Let $\sin\theta = \dfrac{l}{m}$, where $l, m > 0$. Express $\tan\theta$ and $\cos\theta$ in terms of l and m.

This page intentionally left blank

Lesson 31: Using Trigonometry to Determine Area

Classwork

Opening Exercise

Three triangles are presented below. Determine the areas for each triangle, if possible. If it is not possible to find the area with the provided information, describe what is needed in order to determine the area.

 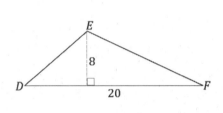 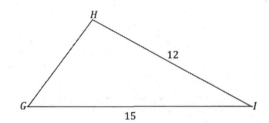

Is there a way to find the missing information?

Example 1

Find the area of △ GHI.

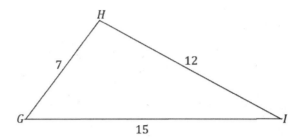

Example 2

A farmer is planning how to divide his land for planting next year's crops. A triangular plot of land is left with two known side lengths measuring 500 m and 1,700 m.

What could the farmer do next in order to find the area of the plot?

Lesson 31 M2
GEOMETRY

Exercise 1

A real estate developer and her surveyor are searching for their next piece of land to build on. They each examine a plot of land in the shape of $\triangle ABC$. The real estate developer measures the length of $\overline{AB}$ and $\overline{AC}$ and finds them to both be approximately 4,000 feet, and the included angle has a measure of approximately 50°. The surveyor measures the length of $\overline{AC}$ and $\overline{BC}$ and finds the lengths to be approximately 4,000 feet and 3,400 feet, respectively, and measures the angle between the two sides to be approximately 65°.

a. Draw a diagram that models the situation, labeling all lengths and angle measures.

b. The real estate developer and surveyor each calculate the area of the plot of land and both find roughly the same area. Show how each person calculated the area; round to the nearest hundred. Redraw the diagram with only the relevant labels for both the real estate agent and surveyor.

c. What could possibly explain the difference between the real estate agent's and surveyor's calculated areas?

Lesson 31: Using Trigonometry to Determine Area

A STORY OF FUNCTIONS   Lesson 31 M2
GEOMETRY

Problem Set

Find the area of each triangle. Round each answer to the nearest tenth.

1.
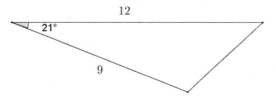

2.

3.
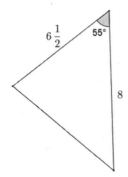

4.
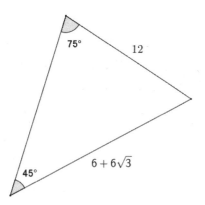

5. In $\triangle DEF$, $EF = 15$, $DF = 20$, and $m\angle F = 63°$. Determine the area of the triangle. Round to the nearest tenth.

6. A landscape designer is designing a flower garden for a triangular area that is bounded on two sides by the client's house and driveway. The length of the edges of the garden along the house and driveway are 18 ft. and 8 ft., respectively, and the edges come together at an angle of 80°. Draw a diagram, and then find the area of the garden to the nearest square foot.

7. A right rectangular pyramid has a square base with sides of length 5. Each lateral face of the pyramid is an isosceles triangle. The angle on each lateral face between the base of the triangle and the adjacent edge is 75°. Find the surface area of the pyramid to the nearest tenth.

8. The Pentagon building in Washington, DC, is built in the shape of a regular pentagon. Each side of the pentagon measures 921 ft. in length. The building has a pentagonal courtyard with the same center. Each wall of the center courtyard has a length of 356 ft. What is the approximate area of the roof of the Pentagon building?

9. A regular hexagon is inscribed in a circle with a radius of 7. Find the perimeter and area of the hexagon.

10. In the figure below, ∠AEB is acute. Show that Area(△ ABC) = $\frac{1}{2}$ AC · BE · sin ∠AEB.

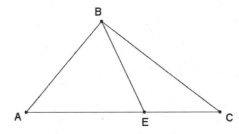

11. Let $ABCD$ be a quadrilateral. Let w be the measure of the acute angle formed by diagonals $\overline{AC}$ and $\overline{BD}$. Show that Area$(ABCD) = \frac{1}{2} AC \cdot BD \cdot \sin w$.

    (Hint: Apply the result from Problem 10 to △ ABC and △ ACD.)

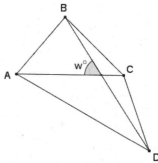

This page intentionally left blank

Lesson 32: Using Trigonometry to Find Side Lengths of an Acute Triangle

Classwork

Opening Exercise

a. Find the lengths of d and e.

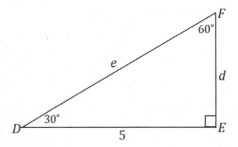

b. Find the lengths of x and y. How is this different from part (a)?

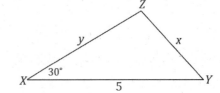

Example 1

A surveyor needs to determine the distance between two points A and B that lie on opposite banks of a river. A point C is chosen 160 meters from point A, on the same side of the river as A. The measures of $\angle BAC$ and $\angle ACB$ are $41°$ and $55°$, respectively. Approximate the distance from A to B to the nearest meter.

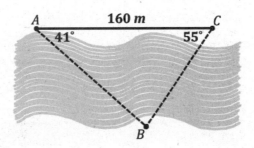

Exercises 1–2

1. In $\triangle ABC$, $m\angle A = 30$, $a = 12$, and $b = 10$. Find $\sin\angle B$. Include a diagram in your answer.

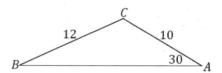

2. A car is moving toward a tunnel carved out of the base of a hill. As the accompanying diagram shows, the top of the hill, H, is sighted from two locations, A and B. The distance between A and B is 250 ft. What is the height, h, of the hill to the nearest foot?

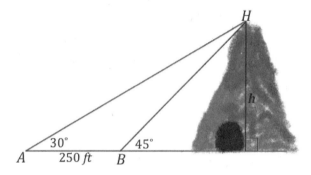

Example 2

Our friend the surveyor from Example 1 is doing some further work. He has already found the distance between points A and B (from Example 1). Now he wants to locate a point D that is equidistant from both A and B and on the same side of the river as A. He has his assistant mark the point D so that $\angle ABD$ and $\angle BAD$ both measure 75°. What is the distance between D and A to the nearest meter?

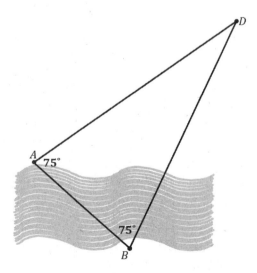

Lesson 32: Using Trigonometry to Find Side Lengths of an Acute Triangle

Exercise 3

3. Parallelogram $ABCD$ has sides of lengths 44 mm and 26 mm, and one of the angles has a measure of 100°. Approximate the length of diagonal $\overline{AC}$ to the nearest millimeter.

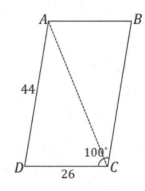

Problem Set

1. Given △ABC, $AB = 14$, $\angle A = 57.2°$, and $\angle C = 78.4°$, calculate the measure of angle B to the nearest tenth of a degree, and use the law of sines to find the lengths of $\overline{AC}$ and $\overline{BC}$ to the nearest tenth.

   Calculate the area of △ABC to the nearest square unit.

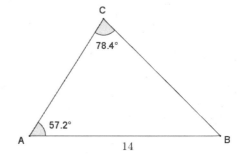

2. Given △DEF, $\angle F = 39°$, and $EF = 13$, calculate the measure of $\angle E$, and use the law of sines to find the lengths of $\overline{DF}$ and $\overline{DE}$ to the nearest hundredth.

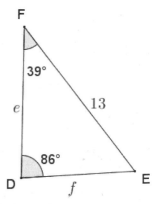

3. Does the law of sines apply to a right triangle? Based on △ABC, the following ratios were set up according to the law of sines.

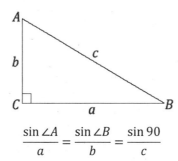

$$\frac{\sin \angle A}{a} = \frac{\sin \angle B}{b} = \frac{\sin 90}{c}$$

Fill in the partially completed work below.

$$\frac{\sin \angle A}{a} = \frac{\sin 90}{c}$$

$$\frac{\sin \angle A}{a} = \frac{}{c}$$

$$\sin \angle A = \frac{}{c}$$

$$\frac{\sin \angle B}{b} = \frac{\sin 90}{c}$$

$$\frac{\sin \angle B}{b} = \frac{}{c}$$

$$\sin \angle B = \frac{}{c}$$

What conclusions can we draw?

4. Given quadrilateral $GHKJ$, $m\angle H = 50°$, $m\angle HKG = 80°$, $m\angle KGJ = 50°$, $m\angle J$ is a right angle, and $GH = 9$ in., use the law of sines to find the length of $\overline{GK}$, and then find the lengths of $\overline{GJ}$ and $\overline{JK}$ to the nearest tenth of an inch.

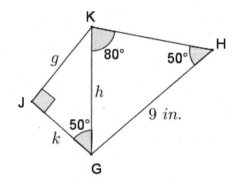

5. Given triangle LMN, $LM = 10$, $LN = 15$, and $m\angle L = 38°$, use the law of cosines to find the length of $\overline{MN}$ to the nearest tenth.

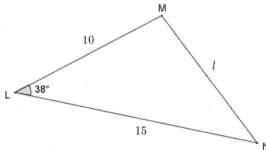

6. Given triangle ABC, $AC = 6$, $AB = 8$, and $m\angle A = 78°$, draw a diagram of triangle ABC, and use the law of cosines to find the length of $\overline{BC}$.

Calculate the area of triangle ABC.

Lesson 32: Using Trigonometry to Find Side Lengths of an Acute Triangle

This page intentionally left blank

Lesson 33: Applying the Laws of Sines and Cosines

Classwork

Opening Exercise

For each triangle shown below, identify the method (Pythagorean theorem, law of sines, law of cosines) you would use to find each length x.

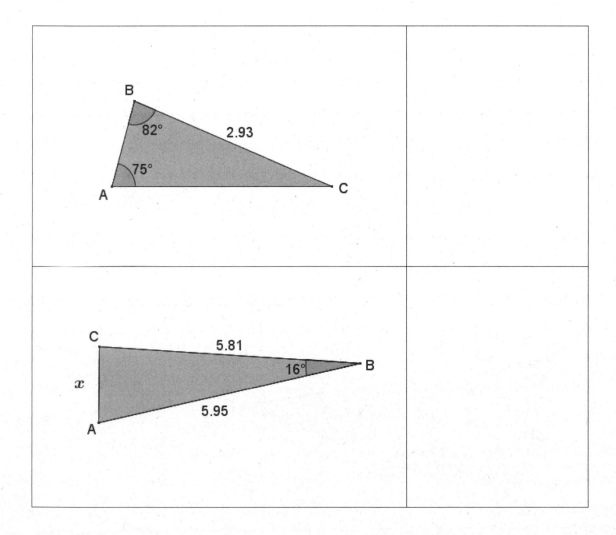

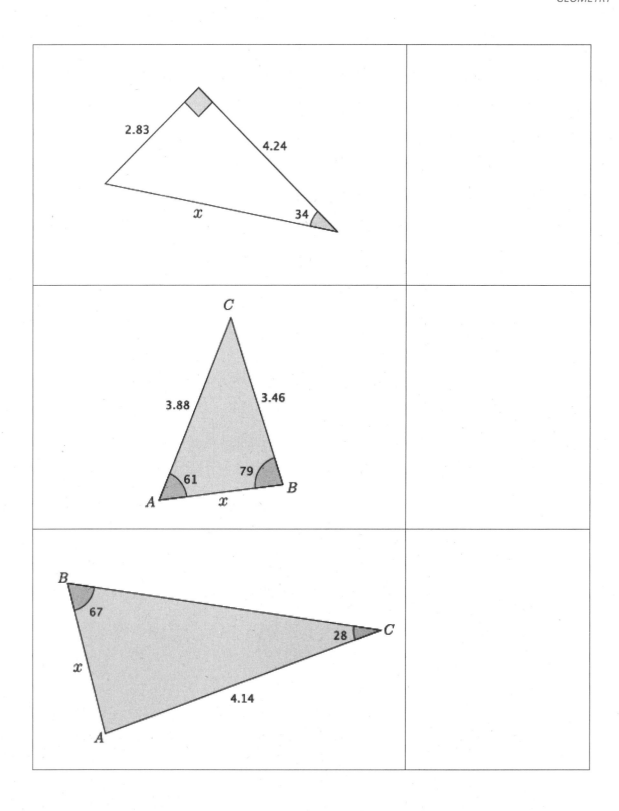

Example 1

Find the missing side length in △ABC.

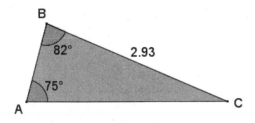

Example 2

Find the missing side length in △ABC.

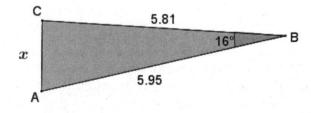

Exercises 1–6

Use the laws of sines and cosines to find all missing side lengths for each of the triangles in the exercises below. Round your answers to the tenths place.

1. Use the triangle to the right to complete this exercise.

   a. Identify the method (Pythagorean theorem, law of sines, law of cosines) you would use to find each of the missing lengths of the triangle. Explain why the other methods cannot be used.

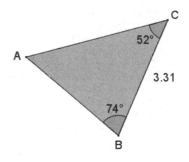

   b. Find the lengths of $\overline{AC}$ and $\overline{AB}$.

2. Your school is challenging classes to compete in a triathlon. The race begins with a swim along the shore and then continues with a bike ride for 4 miles. School officials want the race to end at the place it began, so after the 4-mile bike ride, racers must turn 30° and run 3.5 miles directly back to the starting point. What is the total length of the race? Round your answer to the tenths place.

   a. Identify the method (Pythagorean theorem, law of sines, law of cosines) you would use to find the total length of the race. Explain why the other methods cannot be used.

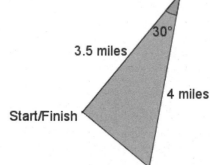

Lesson 33: Applying the Laws of Sines and Cosines

b. Determine the total length of the race. Round your answer to the tenths place.

3. Two lighthouses are 30 miles apart on each side of shorelines running north and south, as shown. Each lighthouse keeper spots a boat in the distance. One lighthouse keeper notes the location of the boat as 40° east of south, and the other lighthouse keeper marks the boat as 32° west of south. What is the distance from the boat to each of the lighthouses at the time it was spotted? Round your answers to the nearest mile.

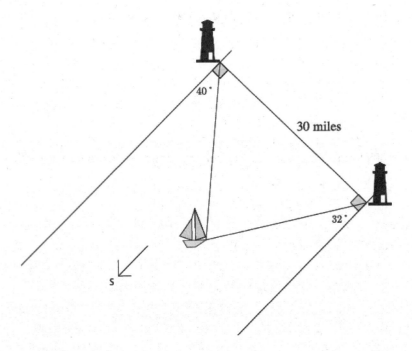

4. A pendulum 18 in. in length swings 72° from right to left. What is the difference between the highest and lowest point of the pendulum? Round your answer to the hundredths place, and explain how you found it.

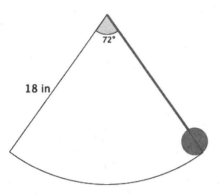

5. What appears to be the minimum amount of information about a triangle that must be given in order to use the law of sines to find an unknown length?

6. What appears to be the minimum amount of information about a triangle that must be given in order to use the law of cosines to find an unknown length?

Problem Set

1. Given triangle EFG, $FG = 15$, angle E has a measure of $38°$, and angle F has a measure of $72°$, find the measures of the remaining sides and angle to the nearest tenth. Justify your method.

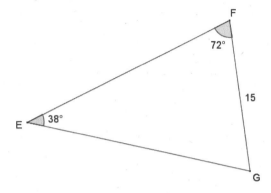

2. Given triangle ABC, angle A has a measure of $75°$, $AC = 15.2$, and $AB = 24$, find BC to the nearest tenth. Justify your method.

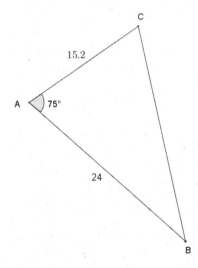

3. James flies his plane from point A at a bearing of $32°$ east of north, averaging a speed of 143 miles per hour for 3 hours, to get to an airfield at point B. He next flies $69°$ west of north at an average speed of 129 miles per hour for 4.5 hours to a different airfield at point C.

   a. Find the distance from A to B.

   b. Find the distance from B to C.

   c. Find the measure of angle ABC.

   d. Find the distance from C to A.

   e. What length of time can James expect the return trip from C to A to take?

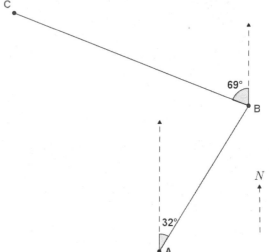

4. Mark is deciding on the best way to get from point A to point B as shown on the map of Crooked Creek to go fishing. He sees that if he stays on the north side of the creek, he would have to walk around a triangular piece of private property (bounded by $\overline{AC}$ and $\overline{BC}$). His other option is to cross the creek at A and take a straight path to B, which he knows to be a distance of 1.2 mi. The second option requires crossing the water, which is too deep for his boots and very cold. Find the difference in distances to help Mark decide which path is his better choice.

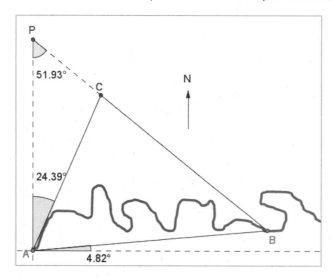

5. If you are given triangle ABC and the measures of two of its angles and two of its sides, would it be appropriate to apply the law of sines or the law of cosines to find the remaining side? Explain.

Lesson 34: Unknown Angles

Classwork

Opening Exercise

a. Dan was walking through a forest when he came upon a sizable tree. Dan estimated he was about 40 meters away from the tree when he measured the angle of elevation between the horizontal and the top of the tree to be 35 degrees. If Dan is about 2 meters tall, about how tall is the tree?

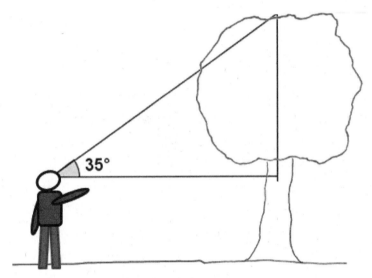

b. Dan was pretty impressed with this tree ... until he turned around and saw a bigger one, also 40 meters away but in the other direction. "Wow," he said. "I bet that tree is *at least* 50 meters tall!" Then, he thought a moment. "Hmm ... if it *is* 50 meters tall, I wonder what angle of elevation I would measure from my eye level to the top of the tree?" What angle will Dan find if the tree is 50 meters tall? Explain your reasoning.

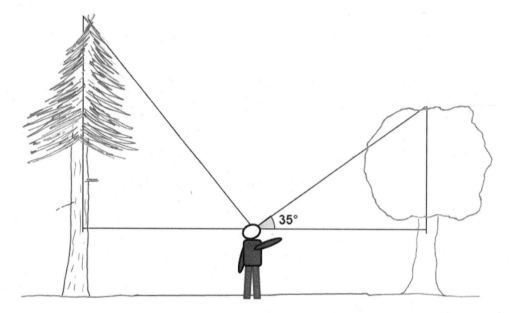

Exercises 1–5

1. Find the measure of angles a through d to the nearest degree.

   a.
   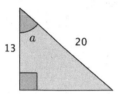

   b.
   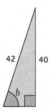

   c.
   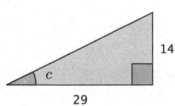

   d.
   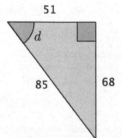

2. Shelves are being built in a classroom to hold textbooks and other supplies. The shelves will extend 10 in. from the wall. Support braces will need to be installed to secure the shelves. The braces will be attached to the end of the shelf and secured 6 in. below the shelf on the wall. What angle measure will the brace and the shelf make?

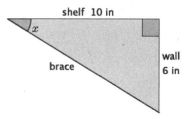

3. A 16 ft. ladder leans against a wall. The foot of the ladder is 7 ft. from the wall.

   a. Find the vertical distance from the ground to the point where the top of the ladder touches the wall.

   b. Determine the measure of the angle formed by the ladder and the ground.

Lesson 34: Unknown Angles

4. A group of friends have hiked to the top of the Mile High Mountain. When they look down, they can see their campsite, which they know is approximately 3 miles from the base of the mountain.

   a. Sketch a drawing of the situation.

   b. What is the angle of depression?

5. A roller coaster travels 80 ft. of track from the loading zone before reaching its peak. The horizontal distance between the loading zone and the base of the peak is 50 ft.

   a. Model the situation using a right triangle.

   b. At what angle is the roller coaster rising according to the model?

Lesson Summary

In the same way that mathematicians have named certain ratios within right triangles, they have also developed terminology for identifying angles in a right triangle, given the ratio of the sides. Mathematicians often use the prefix *arc* to define these because an angle is not just measured as an angle, but also as a length of an *arc* on the unit circle.

Given a right triangle ABC, the measure of angle C can be found in the following ways:

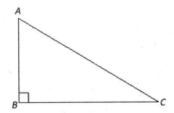

- $\arcsin\left(\dfrac{AB}{AC}\right) = m\angle C$
- $\arccos\left(\dfrac{BC}{AC}\right) = m\angle C$
- $\arctan\left(\dfrac{AB}{BC}\right) = m\angle C$

We can write similar statements to determine the measure of angle A.

We can use a calculator to help us determine the values of arcsin, arccos, and arctan. Most calculators show these buttons as "sin^{-1}," "cos^{-1}," and "tan^{-1}." This subject is addressed again in future courses.

Problem Set

1. For each triangle shown, use the given information to find the indicated angle to the nearest degree.

   a.

   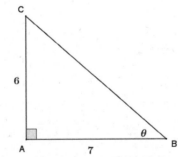

   b.

   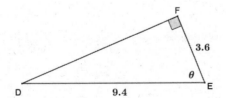

c.

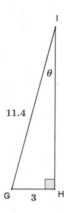

2. *Solving a right triangle* means using given information to find all the angles and side lengths of the triangle. Use arcsin and arccos, along with the given information, to solve right triangle ABC if leg $AC = 12$ and hypotenuse $AB = 15$.

   Once you have found the measure of one of the acute angles in the right triangle, can you find the measure of the other acute angle using a different method from those used in this lesson? Explain.

3. A pendulum consists of a spherical weight suspended at the end of a string whose other end is anchored at a pivot point P. The distance from P to the center of the pendulum's sphere, B, is 6 inches. The weight is held so that the string is taut and horizontal, as shown to the right, and then dropped.

   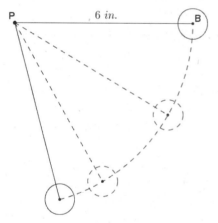

   a. What type of path does the pendulum's weight take as it swings?

   b. Danni thinks that for every vertical drop of 1 inch that the pendulum's weight makes, the degree of rotation is 15°. Do you agree or disagree with Danni? As part of your explanation, calculate the degree of rotation for every vertical drop of 1 inch from 1 inch to 6 inches.

4. A stone tower was built on unstable ground, and the soil beneath it settled under its weight, causing the tower to lean. The cylindrical tower has a diameter of 17 meters. The height of the tower on the low side measured 46.3 meters and on the high side measured 47.1 meters. To the nearest tenth of a degree, find the angle that the tower has leaned from its original vertical position.

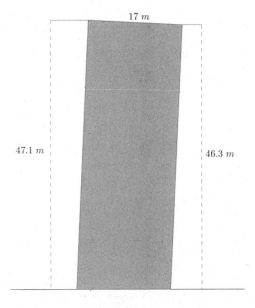

5. Doug is installing a surveillance camera inside a convenience store. He mounts the camera 8 ft. above the ground and 15 ft. horizontally from the store's entrance. The camera is being installed to monitor every customer who enters and exits the store. At what angle of depression should Doug set the camera to capture the faces of all customers?

Note: This is a modeling problem and therefore will have various reasonable answers.

Lesson 34: Unknown Angles

This page intentionally left blank

This page intentionally left blank

This page intentionally left blank

This page intentionally left blank